MYELOID CELLS

BIOLOGY & REGULATION, ROLE IN CANCER PROGRESSION AND POTENTIAL IMPLICATIONS FOR THERAPY

CELL BIOLOGY RESEARCH PROGRESS

Additional books in this series can be found on Nova's website
under the Series tab.

Additional e-books in this series can be found on Nova's website
under the e-book tab.

CELL BIOLOGY RESEARCH PROGRESS

MYELOID CELLS

BIOLOGY & REGULATION, ROLE IN CANCER PROGRESSION AND POTENTIAL IMPLICATIONS FOR THERAPY

SPENCER A. DOUGLAS
EDITOR

New York

Library of Congress Cataloging-in-Publication Data

ISBN: 978-1-62948-046-6

Library of Congress Control Number: 2013948185

Published by Nova Science Publishers, Inc. † New York

CONTENTS

PREFACE

In this book the authors present current research in the study of the biology and regulation, role in cancer progression and potential implications for therapy of myeloid cells. Topics discussed include the differentiation signaling induced by retinoic acid and vitamin D3; proline rich homeodomain (Prh/Hhex) protein in the control of haematopoiesis and myeloid cell proliferation and its potential as a therapeutic target in myeloid leukaemias and other cancers; apoptosis, cell cycle and epigenetic processes deregulation in myeloproliferative neoplasms; use of animal models to evaluate myeloid cell dysfunction in cancer; use of animal models to evaluate myeloid cell dysfunction in cancer; and the biology of myeloid cells mediating tumor recurrence after radiotherapy.

Chapter 1 - All-trans retinoic acid (ATRA) and 1, 25-dihydroxyvitamin D3 $(1,25(OH)_2D_3)$ are involved in a variety of important physiological processes, among which their role in the induction of differentiation of leukemia cells has attracted particular attention in recent years. Acute myeloid leukemia (AML) cells can be stimulated to differentiate to granulocyte-like cells by ATRA and to monocytes by $1,25(OH)_2D_3$. The biological effects of these two molecules are mainly mediated by the nuclear retinoid acid (RAR) and vitamin D receptors (VDR), respectively. In each case, following ligand binding, the RARs and VDR form a heterodimer complex with retinoid X receptors (RXRs). These heterodimeric complexes regulate DNA transcription by binding to the promoter regions of various target genes in the presence of co-activators. In the absence of ligand, RAR-RXR or VDR-RXA heterodimers bind to corepressors and are not capable of inducing transcription of these target genes. Acute promyelocytic leukemia (APL) is a rare subtype of AML, which is characterized by the presence of an abnormal PML-RAR fusion protein. ATRA is able to disrupt the PML/RARα-RXR complex and restore the RARα-RXR signaling. Several transcription factors, such as FOXO3A and c-Myc, have been reported to participate in ATRA-induced differentiation in AML and other blood cells and many ATRA and $1,25(OH)_2D_3$ target genes including UBE1L, UBE2D3, C/EBPs, TRAIL, FOXOs, CAMPs, OLFM4, and c-Myc, have been identified in recent years. There is now accumulating evidence that a number of cell cycle regulatory molecules and multiple intracellular signal pathways/molecules, such as Raf-ERK, PKC isoforms, and PI3-K-AKT, may also regulate ATRA and $1,25(OH)_2D_3$ induced differentiation of the leukemia cells. However, many questions remain to be answered: What is the degree of cross-talk between genomic regulation and intracellular signaling pathways? Which gene(s) or signal molecule(s) play a key role in determining the differentiation of AML cells into granulocytes or monocytes?

Chapter 2 - Myelopoiesis occurs as a consequence of the interplay between transcription factors and growth factor dependent signalling pathways impacting on cell proliferation, cell survival and cell differentiation. It is equally well established that mis-regulation of transcription factors and/or growth factor receptor pathways contributes to disruption of myelopoeisis and the genesis of chronic and acute myeloid leukaemias. The Proline Rich Homeodomain protein (PRH, also known as HHEX) is a transcription factor that is essential for haematopoiesis and it is expressed in haematopoietic stem cells and all myeloid lineages. PRH also plays a role in early embryonic development and in the formation of many organ systems. Significantly, there is compelling evidence that PRH is mis-regulated in both chronic myeloid leukaemia and some subtypes of acute myeloid leukaemia as well as in T-cell leukaemias and some solid cancers. This chapter reviews the functions of PRH in embryonic development, normal haematopoiesis and in myeloid leukaemias and focuses on the structure, localisation, transcriptional activity and post-translational modifications of PRH. In myeloid cells, PRH has been shown to be a transcriptional repressor of genes that regulate cell proliferation/survival including multiple genes in the VEGF signalling pathway. Thus PRH is a potent inhibitor of myeloid cell proliferation. Protein kinase CK2 is a stress responsive kinase with pleiotropic activities that promotes cell proliferation and its activity is often elevated in leukaemia and solid tumours. Phosphorylation of PRH by CK2 inhibits DNA binding by PRH and alleviates growth inhibition by PRH. Therapeutic treatment for CML (imatinib and dasatinib) also influences PRH phosphorylation and increases transcriptional repression of growth regulatory genes by PRH. Here the authors review the evidence which suggests that PRH may be an important therapeutic target in CML and other leukaemias, as well as in other cancers.

Chapter 3 - Myeloproliferative neoplasms (MPN) are hematological diseases characterized by myeloproliferation/myeloaccumulation of mature cells without a specific stimulus. According to the World Health Organization (WHO, 2008), MPN are composed by Philadelphia (Ph) chromosome negative MPN, including Primary Myelofibrosis (PMF), Essential Thrombocythemia (ET), Polycythemia Vera (PV), uncommon MPN as well as chronic neutrophilic leukemia, mastocytosis and by Chronic Myeloid Leukemia (CML), a Ph positive chromosome MPN. In this chapter the authors focus on Primary Myelofibrosis (PMF), Essential Thrombocythemia (ET), Polycythemia Vera (PV) and Chronic Myeloid Leukemia (CML) pathogenesis, specifically apoptosis, cell cycle and epigenetic processes deregulation.

The JAK2 V617F mutation, which leads to constitutive JAK2 tyrosine kinase activation, is found in 95% of PV patients and in at least 50% of ET and PMF patients. JAK2 constitutive enzymatic activation is linked to prolonged cell survival and myeloproliferation. Other mutations, recently described, such as in JAK2 exon 12 and in the TET2, CBL, MPL and AXSL genes may also contribute to ET, PV and PMF pathogenesis and progression to acute myeloid leukemia. Some acquired genetic lesions in molecules involved in cytokine signaling and epigenetic regulation are also implicated in Ph negative MPN.

In CML patients, the expression of BCR-ABL1 determines the leukemogenesis process by increasing cell proliferation, promoting apoptosis impairment and deregulation of cell adhesion to bone marrow stroma and altering the cell epigenetic pattern.

Elucidation of cellular and molecular mechanisms involved in MPN pathogenesis is relevant for new therapeutic targets discovery and for description of potentials diagnostic and prognostic markers for diseases.

Chapter 4 - In recent years, the concept of using the body's own immune system to target and kill tumor cells has been an appealing approach to treat cancer. Cancer immunotherapy holds particular promise for treating metastatic disease but has met with relatively limited success due to our limited understanding of how the immune system is dysregulated in cancer. Myeloid cells are a diverse population of immune cells that are markedly altered in cancer. Comprised of macrophages, dendritic cells (DCs) and myeloid-derived suppressor cells (MDSCs), these significantly contribute to the generation of antitumor immune responses and influence the development and progression of cancer. In order to better understand the role of myeloid cells in the tumor microenvironment, it is necessary to develop and characterize various cancer models. In this review, the authors will first outline how cancer creates dysfunction in various myeloid cell populations. While evaluating human cells and tissue is ideal for cancer research, there are many limitations to what questions can be investigated. To fill these gaps, there have been several successful murine and canine models that are comparative oncology models to study cancer immunology. To understand the potential application of animal models in cancer immunology, this review will also outline some of the benefits and limitations of current animal models and discuss their relevance to studying myeloid cell dysfunction in cancer. Given the importance of myeloid cells in antitumor immune responses and developing efficacious cancer immunotherapies, a better understanding of myeloid cell dysfunction in cancer is necessary, regardless of species.

Chapter 5 - cancer that bone marrow-derived myelomonocytic cells infiltrate into tumors and play a critical role in promoting tumor recurrence after radiotherapy. Critical attributes of these bone marrow-derived myelomonocytic cells are their highly proangiogenic nature and the expression of matrix metalloproteinase-9 (MMP-9) and the CXCR4 chemokine receptor, which responds to stromal-derived factor-1 (SDF-1) produced by irradiated tumors. The recruited myelomonocytic cells in the irradiated tumors then support immature blood vessel development thereby promoting re-growth of the tumor in the irradiated vascular bed. In this review, the authors focus on some of the signaling pathways occurring between irradiated tumors and recruited myelomonocytic cells, including MMP-9, the SDF-1-CXCR4 axis, tumor necrosis factor alpha (TNF-α), and colony-stimulating factor-receptor (CSF-R) activation pathways, each of which had been reported to promote tumor re-growth after irradiation. The authors will also discuss clinically relevant strategies to inhibit myelomonocytic influx into the irradiated tumors as well as issues to be considered when these strategies are to be translated into the clinic.

In: Myeloid Cells
Editor: Spencer A. Douglas

ISBN: 978-1-62948-046-6
© 2013 Nova Science Publishers, Inc.

Chapter 1

DIFFERENTIATION SIGNALING INDUCED BY RETINOIC ACID AND VITAMIN D3

Xiaotang Hu[*1] *and Graham Shaw*[2]
[1]Department of Biology, College of Arts & Sciences,
[2]The School of Podiatric Medicine, Barry University,
Miami Shores, FL

ABSTRACT

All-trans retinoic acid (ATRA) and 1, 25-dihydroxyvitamin D3 $(1,25(OH)_2D_3)$ are involved in a variety of important physiological processes, among which their role in the induction of differentiation of leukemia cells has attracted particular attention in recent years. Acute myeloid leukemia (AML) cells can be stimulated to differentiate to granulocyte-like cells by ATRA and to monocytes by $1,25(OH)_2D_3$. The biological effects of these two molecules are mainly mediated by the nuclear retinoid acid (RAR) and vitamin D receptors (VDR), respectively. In each case, following ligand binding, the RARs and VDR form a heterodimer complex with retinoid X receptors (RXRs). These heterodimeric complexes regulate DNA transcription by binding to the promoter regions of various target genes in the presence of co-activators. In the absence of ligand, RAR-RXR or VDR-RXA heterodimers bind to

* Address all correspondence to: Xiaotang Hu, Ph.D. Professor of Biology, Department of Biology, College of Arts & Sciences, Barry University, 11300 Northeast Second Avenue, Miami Shore, FL 33161, Phone: (305)899-3295, Fax: (305)899-3225, E-mail: xthu@mail.barry.edu.

corepressors and are not capable of inducing transcription of these target genes. Acute promyelocytic leukemia (APL) is a rare subtype of AML, which is characterized by the presence of an abnormal PML-RAR fusion protein. ATRA is able to disrupt the PML/RARα-RXR complex and restore the RARα-RXR signaling. Several transcription factors, such as FOXO3A and c-Myc, have been reported to participate in ATRA-induced differentiation in AML and other blood cells and many ATRA and $1,25(OH)_2D_3$ target genes including UBE1L, UBE2D3, C/EBPs, TRAIL, FOXOs, CAMPs, OLFM4, and c-Myc, have been identified in recent years. There is now accumulating evidence that a number of cell cycle regulatory molecules and multiple intracellular signal pathways/molecules, such as Raf-ERK, PKC isoforms, and PI3-K-AKT, may also regulate ATRA and $1,25(OH)_2D_3$ induced differentiation of the leukemia cells. However, many questions remain to be answered: What is the degree of cross-talk between genomic regulation and intracellular signaling pathways? Which gene(s) or signal molecule(s) play a key role in determining the differentiation of AML cells into granulocytes or monocytes?

INTRODUCTION

Hematopoiesis is the formation and development of blood cells, including cell proliferation, differentiation, and cell death. All mature blood cells originate from a rare population of primitive pluripotent (multipotential) stem cells. These pluripotent stem cells possess self-renewal and multi-lineage differentiation potential that leads to two major multipotent progenitors: myeloid progenitors and lymphoid progenitors. The myeloid progenitors have the ability to differentiate into erythrocytes, megakaryocytes/platelets, mast cells, and mature myeloid cells; whereas the lymphoid progenitor cells mainly produce mature T, NK, and B lymphocytes (Figure 1). Under normal circumstances, proliferation, differentiation and cell death are tightly regulated. Over proliferation of immature cells could bring about leukemia while abnormal differentiation or apoptosis could lead to severe cytopenias. Upregulation of differentiation is usually associated with downregulation of proliferative capacity, i.e., the cells quit the cell cycle and enter the G1/G0 phase. In adults, both the cells and extracellular environment (ECM) express or release a variety of factors that can bind to specific receptors located either at the cell surface or within cytosol and the nucleus.

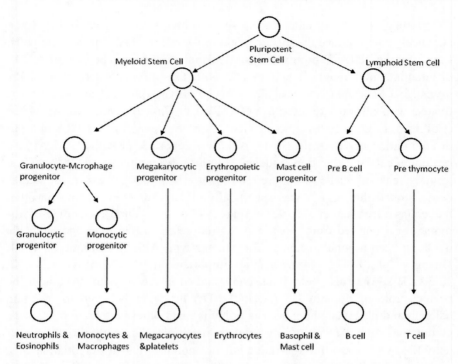

Figure 1. Diagram of Hematopoiesis in Human. Granulocyte-Macrophage Progenitors can differentiate into neutrophils/eosinophils and monocytes/macrophages.

Activation of these receptors transduces the signals to the nuclear DNA and induces the DNA transcription and subsequent translation. The interactions between the molecules involved in the intracellular signal network and the nuclear receptors also play important role in the balance between the proliferation and the differentiation status of the cells.

Myeloid leukemia is a type of blood and bone marrow cancers, which includes acute myeloid leukemia (AML) and chronic myelogenous leukemia (CML). A characteristic abnormality of AML leukemia cells is that they are blocked at an early stage of their development and fail to differentiate into functional mature cells. Acute promyelocytic leukemia (APL) is a subtype of AML characterized by the t(15;17) chromosomal translocation and the expression of an abnormal PML-RARα fusion protein (Pandolfi et al., 1991; de Thé et al., 1991; Kakizuka et al., 1991; Kastner et al. 1992). The abnormal PML-RARα complex blocks leukemia cell differentiation and causes accumulation of immature leukemia cells (see below). Thus, induction of cell differentiation is a major strategy for anti-AML and other cancer therapy.

During the 1970s and 1980s, several investigators found that some chemicals could induce the differentiation of leukemia cells into mature cells. For example, HL-60 promyelocytic leukemia cells can be stimulated to differentiate into monocytes if treated with the phorbol ester, phorbol-12-myristate-13-acetate (Rovera et al., 1979), or polymorphonuclear leukocytes if treated with all-trans retinoic acid (ATRA) or dimethyl sulfoxide (DMSO) (Collins et al., 1978). Soon, the first evidence of ATRA-induced differentiation of leukemic HL-60 cells was reported, in which ATRA, at a physiological concentration, induced terminal differentiation in 90% of the primary leukemia cells in culture (Breitman et al., 1980). Subsequent studies demonstrated that ATRA was specifically effective in APL but not in other leukemias (Breitman et al., 1981). Since ATRA is a natural metabolite of all-trans-retinal derived from vitamin A that is present in the food, the molecule has long been a popular therapy for patients with APL (Nilsson et al., 1984; Huang et al., 1987, 1988). Another important metabolite of vitamin D3, $1,25(OH)_2D_3$, has also been intensively studied in recent years. Although the primary role of vitamin D3 $(1,25(OH)_2D_3)$ has long believed to maintain calcium and phosphate homeostasis in human and other vertebrate organisms, cumulative studies *in vitro* now suggest that it also has multiple anti-cancer activities, which are through induction of apoptosis and differentiation of a variety of different types of malignant cells including leukemia cells. In 1981 a group of investigators found that mouse M1 myeloid cells could be induced to macrophages by $1,25(OH)_2D_3$ (Abe et al., 1981). Two years later, this role of $1,25(OH)_2D_3$ in the induction of differentiation was also observed in mouse leukemic cells (Honma et al., 1983). Since then, vitamin D3-induced differentiation has been observed in various types of human acute myeloid leukemia cells including HL-60 (Mangelsdorf et al., 1984; Brown et al., 1994; Wang and Studzinski 2001), U937 (Tanaka et al., 1992; Botling et al., 1996; Bobilev et al., 2011), NB4 (Hughes et al., 2005), THP-1 (Hmama et al., 1999; Jamshidi et al., 2008), and KG-1 cells (Gocek et al., 2012).

Although the biological effects of ATRA and $1,25(OH)_2D_3$ have been well documented and both molecules have been used in clinical trials as anti-leukemia and -other cancer therapy, the molecular mechanisms responsible for the induction of differentiation are far from fully understood. There is emerging evidence that the biological effects of these two molecules on cell differentiation are achieved at the genomic level and by interactions between their nuclear receptors and a number of intracellular signals derived from the activation of plasma receptors.

This review summarizes our current understanding of the differentiation signals activated by ATRA and $1,25(OH)_2D_3$ with a focus on cell cycle regulatory molecules and intracellular signal pathways. We apologize to authors whose work could not be cited due to page limitation concerns.

GENOMIC CONTROL

1. Nuclear Receptors

It has long been recognized that the biological effects of ATRA and $1,25(OH)_2D_3$ are mainly mediated by the retinoic acid receptor (RAR) and vitamin D receptor (VDR), respectively. These receptors belong to the type II class of nuclear receptors (NRs) that are usually retained in the nucleus regardless of presence or absence of their ligand. However, ligand binding is required for DNA transcription activation (Mangelsdorf et al., 1995; Krauss et al., 2001; Rhinn and Dollé 2012). In the absence of a ligand, type II receptors are often complexed with corepressor proteins. However, a number of studies have showed that RAR and VDR are able to shuttle between the nucleus and cytoplasm without a ligand, but binding of ligand promotes these heterodimers nuclear localization (Zhu et al., 1998; Racz and Barsonyh 1999; Prüfer and Barsony 2002, Capiati et al., 2004; Yasmin et al., 2005). Interestingly, significant quantities of liganded-VDR have also been found in the cytoplasm (Kim et al., 1996). All nuclear receptors share a common structure with five domains/regions: a N-terminal regulatory domain (A/B domain), a highly conserved DNA binding domain (DBD), a Hinge region (H region), a ligand binding domain (LBD), and a C-terminal domain (Tata 2002; Robinson-Rechavi et al., 2003). Both the A/B domain and the c-terminal domain vary highly in sequence with the type of nuclear receptors. The ligand-binding domain provides a binding site for cognate ligand. The hinge region, which is located between DBD and LBD, contains the nuclear localization signal and is a target site for methylation, acetylation and ubiquitination. It has been reported that phosphorylation of the hinge is coupled with upregulation of transcription activation (Lee et al., 2006). The DBD has two zinc fingers that can bind to hormone response elements (HRE) in the promoter region of the target gene.

Upon ATRA binding, the RAR form heterodimer with the retinoid X receptor (RXR), another member of the nuclear hormone receptor family. Previous studies have demonstrated that ATRA only binds to RAR but not

RXR, whereas 9-cis-retinoic acid (9-cis-RA), a related metabolite of ATRA, is capable of binding to both RAR and RXR (Levin et al., 1992; Heyman et al., 1992). In order to initiate DNA transcription, several co-activators are recruited to the heterodimer (RAR-RXR or VDR-RXR) and DNA biding site. In humans, there are three types of RAR and RXR: α, β, and γ. The genes encoding RARα, RARβ, RARγ, are located on chromosomes 17, 3, and 12, respectively (Mattei et al., 1991); the genes for RXRα, RXRβ and RXRγ are found on chromosomes 9, 6, and 1, respectively (Almasan et al. 1994). The multiple isoforms of RARs and RXRs may explain, at least in part, why ATRA has pleiotropic biological effects.

However, ATRA-induced granulocytic differentiation of HL-60 cells is mediated primarily through the RARα (Collins et al., 1990; Robertson et al., 1992). In APL patients the presence of an abnormal PML-RARα fusion protein is directly linked to the disease (de Thé et al., 1991; Pandolti P 1991).

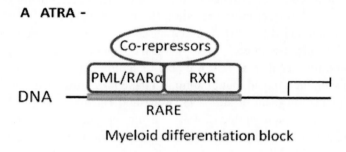

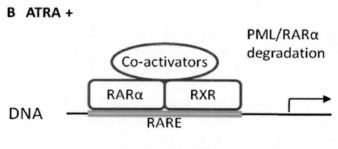

Figure 2. ATRA degrades PML/RARα and induces target gene transcription, resulting in cell differentiation.

The PML-RARα/RXR complex inhibits gene transcription and blocks the differentiation of leukemia cells at the promyelocyte stage, leading to accumulation of the leukemic cells (Zelent et al., 2001; Salomoni and Pandolfi 2002; Ricote et al., 2006). One of the major mechanisms responsible for the anti-leukemia effect of ATRA is due to its ability to disrupt the PML/RARα-RXR complex and restore the RARα-RXR signaling in APL cells (Pitha-Rowe et al., 2003; Jing et al., 2003) (Figure 2). *In vitro*, ATRA and its metabolites degraded PML-RARα and induced leukemic NB4 cell differentiation (Idres et al., 2001).

Compared with RXRβ and RXRγ, the RXRα may play a key role in ATRA and 1,25(OH)2D3-induced differentiation of leukemia cells, which is consistent with the observation that RXRα is predominantly expressed in hematopoietic cells, whereas RXRγ is mostly found in muscles and brain tissue. The RXRβ appears ubiquitously expressed (Ricote et al., 2006; Brown et al., *1997;* Fritsche et al., 2000; Ohata et al., 2000).

As previously described for RAR, the vitamin D receptor (VDR) also needs to bind to RXR to form a heterodimer. This is followed by causing conformational changes that allow the heterodimer binding to vitamin D response elements (VDREs) in the promoter region of target genes. The heterodimer then recruits coactivators, such as vitamin D receptor-interacting protein complex (DRIP) (Rachez et al., 1999), histone acetylase and other proteins. As a result of the interactions of these molecules, DNA is now accessible to transcription factors and transcription is activated. In the absence of Vitamin D3 (1, 25(OH)$_2$D$_3$), the VDR-RXR heterodimer binds to corepressors, recruiting histone deacetylases (HDACs), and preventing the opening of chromatin, resulting in transcriptional repression.

2. Target Genes

2.1. ATRA Target Genes

Several ATRA target genes are involved in the ubiquitin-proteasome system. Degradation of cellular proteins plays a major role in a variety of cellular pathways during both cell living and death, and hence in both health and disease. Degradation of proteins is highly controlled process regulated by the ubiquitin-proteasome system. Modification of proteins by ubiquitins is the first step for the protein degradation by the proteasome. There are three classes of ubiquitin enzymes: ubiquitin-activating enzyme E1; ubiquitin-conjugating enzyme E2; and ubiquitin-protein ligase E3. The first ATRA target gene identified is ubiquitin-like modifier-activating enzyme 7 (UBE1L). UBE1L is encoded by UBA7 gene and has ubiquitin E1 enzyme activity (Kok et al., 1993). The enzyme triggers degradation of PML-RARα/RXR proteins and apoptosis in acute promyelocytic anemia cells (Kitareewan et al., 2002). Another ubiquitin gene targeted by ATRA is UBE2D3. This gene belongs to the ubquitin-conjugating enzyme E2 family. UBE2D3 enzyme is upregulated and is physically associated with cyclin D1 in ATRA-treated acute promyelocytic NB4 cells, which is correlated with ATRA-induced cell arrest. In contrast, knocking out the UBE2D3 by RNA interference (RNAi) blocks ATRA-induced cyclin D1 degradation and cell-cycle arrest (Hattori et al., 2007). The CCAAT-enhancer-binding proteins (C/EBPs) are a family of transcription factors with at least six members: α, β, γ, δ, ε, and ζ. ATRA-induced differentiation of NB4 APL cells is correlated with expression of C/EBPε (Truong et al., 2003) and C/EBPβ (Duprez et al., 2003). In U937 cells, forced expression of C/EBPε caused significant granulocytic differentiation (Park et al., 1999). In contrast, the inhibition of C/EBPβ by using a dominant-negative C/EBP and small interfering RNA (siRNA) against C/EBPb, reduced ATRA-induced maturation of APL cells (Duprez et al., 2003). In addition, C/EBPα has also been reported to have a positive effect on differentiation of hematopoietic cells and enhance ATRA-induced differentiation and growth inhibition of AML cells (Kandilci et al., 2009; Fujiki et al., 2012; Yoshida et al., 2012).

It has been observed that ATRA induces differentiation accompanied by cell death in APL NB4 cells, suggesting that ATRA may also activate apoptotic pathway. One of apoptotic target genes by ATRA is tumor necrosis factor-related apoptotosis-inducing ligand (TRAIL). This protein is produced and secreted by most normal tissue cells. TRAIL acts as a ligand and the ligand binding induces apoptosis primarily in cancer cells. Treatment of NB4 APL cells with RA (or ATRA) induces significant granulocytic differentiation as well as apoptosis of the tumor cells. The RA-induced cell death is regulated by ACTR-activated TRAIL cascade (Altucci et al., 2001; Dhandapani et al., 2001).

Forkhead box O (FOXO) is a group of transcription factors that belong to the FOX superfamily (Tuteja and Kaestner 2007a, 2007b). These transcription factors possess tumor suppressor functions by regulating expression of the genes involved in cell death and proliferation. For example, upregulation of FOXO activates proapoptotic genes encoding for Fas ligand, Bim and TRAIL (Brunet et al., 1999, Dijkers et al., 2000, Modur et al., 2002). FOXO proteins also arrest cells in G1 by upregulating CDK inhibitors, p27 (Medema et al.,

2000, Nakamura et al., 2000), or p130-E2F4 complexes (Kops et al., 2002). The FOXO subfamily includes FOXO1, FOXO3A, FOXO4 and FOX6 in humans. The role of FOXO3A in hemotopoiesis has attracted particular attention in recent years (Miyamoto et al., 2007; Tothova et al., 2007). Phosphorylated FOXO3a is associated with 14-3-3 protein as an inactive form that is mainly found in the cytosol (Komatsu et al., 2003; Birkenkamp and Coffer 2003). The phosphorylation of FOXO3 is primarily regulated by AKT, a serine/threonine kinase (Woods and Rena 2002). In contrast, the dephosphorylated FOXO3a is an active form and functions as a transcription factor. Recently, it was demonstrated that FOXO3a was constitutively phosphorylated and localized to the cytosol in primary APL cells and NB4 cell line. The cells treated with ATRA showed a reduced phosphorylation of FOXO3a that was largely found in the nucleus. Transfection with short hairpin RNA (shRNA) oligonucleotide specific to FOXO3a inhibited ATRA-induced granulocytic differentiation. Furthermore, the ATRA resistant NB4/RA cells did not show any FOXO3a phosphorylation, nuclear localization and differentiation in response to ATRA treatment. These data suggest that ATRA-induced differentiation and apoptosis is through activation of FOXO3a transctiption factor (Sakoe et al., 2010). Another important ATRA target is Olfactomedin 4 (OLFM4) that is a myeloid-lineage–specific gene. This gene was first reported to be activated by granulocyte colony-stimulating factor (G-CSF) and linked to myeloid lineage differentiation (Zhang et al., 2002). Recently, Liu and his colleagues demonstrated that ATRA-induced RAR-RXR binds to the OLFM4 promoter and activates transactivation of the OLFM4 gene. *In vitro,* OLFM4 overexpression in HL-60 cells led to growth inhibition, differentiation, and apoptosis, and enhanced ATRA induction of these effects. In contrast, down-regulation of endogenous OLFM4 in acute myeloid leukemia-193 cells compromised ATRA-induced growth inhibition, differentiation, and apoptosis (Liu et al., 2010).

By using DNA array and several other approaches, one group detected 169 RA target genes in APL leukemic NB4 cells, among which 100 genes are upregulated and 69 genes are downregulated by ATRA (Liu et al., 2000). Whether and how the changes in the gene expression are directly linked to APL leukemogenesis or ATRA-induced differentiation is unclear.

2.2. 1,25(OH)$_2$D$_3$ Target Genes

C/EBP is targeted not only by ATRA but also by 1,25(OH)$_2$D$_3$. The C/EBP family proteins have at least six members: α, β, γ, δ, ζ, and ε. Although both C/EBPα (Friedman et al., 2007) and C/EBPβ are required for myeloid differentiation (Nagamura-Inoue et al., 2001, Ji and Studzinski 2004, Hackanson et al., 2008), C/EBPβ plays a crucial role in $1,25(OH)_2D_3$-induced differentiation (Hackanson et al., 2008). A number of studies have shown that the expression of C/EBPα is only transiently increased during the early stage of the differentiation, whereas the upregulation of C/EBPβ is much stronger and prolonged and directly correlated with the whole differentiation process (Marcinkowska et al., 2006). The key role of C/EBPβ in the cell differentiation process was also demonstrated by the observation that the majority of upregulated C/EBPα was found in the cytosol, whereas the most activated C/EBPβ was translocated from the cytosol into the nucleus in response to $1,25(OH)2D3$ (Marcinkowska et al., 2006). However, in breast cancer cell line, MCF-7, C/EBPα appears to have a profound anti-proliferation effect and is a good candidate for breast cancer treatment (Dhawan et al., 2009). These data suggest that the functions of C/EBP family members may be cell type dependent.

Cathelicidin antimicrobial peptides (CAMPs) are found in the lysosomes of polymorphonuclear leukocytes, macrophages, and epithelial cells and serve a critical role in mammalian innate immune defense against invasive bacterial infection (Nizet et al., 2001). It has long been believed that the synthesis of CAMPs is enhanced in response to stimulation by bacteria, viruses, fungi, and parasites. In 2005 a group of investigators reported that after exposure of human AML leukemia cells including U937, HL60, and NB4 to $1,25(OH)_2D_3$, CAMP gene was significantly upregulated detected by QRT-PCR (Gombart et al., 2005). In U937 cells CAMP induction began between 1 and 3 h after addition of $1,25(OH)_2D_3$ and prior to induction of the differentiation marker CDllb at 12 h. The induction of CAMP gene by $1,25(OH)_2D_3$ is very specific, because application of many other inducers including inflammatory factors (LPS, TPA, TNF-α, INF-α, and INF-γ), cytokines/growth factors (IL-2, IL-6, GM-CSF and G-CSF) and steroid hormones (DHT, estradiol, and ATRA) failed to induce CAMP transcription. However, $1,25(OH)_2D_3$-induced expression of CAMP gene is not specific for granulocytic differentiation, because this induction was also observed in immortalized keratinocytes, and colon cancer cell lines, as well as normal human bone marrow (BM)-derived macrophages and fresh BM cells from two normal individuals (Gombart et al., 2005).

3. Cell Cycle Regulatory Molecules

Involvement of cell cycle regulatory molecules in signaling of granulocytic or monocytic differentiation induced by ATRA or $1,25(OH)_2D_3$ appears at both transcription-dependent and -independent levels. All growing cells undergo a cell cycle. In mammalian cells the cell cycle is divided into four distinct phases: G1, S, G2 and M phases. Viable cells may enter a resting state called G0 phase. The cells in the G0 phases either have irreversibly exited the cell cycle for a particular process or function, such as differentiation, or return to G1 with appropriate stimulation. The cell cycle is controlled by a group of proteins termed cyclin-dependent kinases (CDKs) (Nurse. 2000; Roberts. 1999). CDKs only become active when bound to a regulatory subunit called cyclin (Roberts. 1999). The cell cycle progress through G2 into M phase is driven by cdc2 (CDK1) complexed with cyclin B (also termed "G2 checkpoint kinase"). The G1 cyclin-cdk complexes (also termed "G1 checkpoint kinases") regulate the progress of the cells through G1 toward DNA replication (S phase). There are at least five major G1 cyclins, termed cyclins D1, D2, D3, A, and E. Each of these cyclins can associate with one or more of the G1 CDK family (CDK2, CDK4, and CDK6). CDK/cyclin complexes are generally considered to act on DNA replication machinery in the nucleus, however, Cyclin A- and Cyclin E-Cdk complexes have been found to shuttle between the nucleus and the cytoplasm (Jackman et al., 2002), suggesting that these CDKs/cyclins may also function in the cytoplasm. Full activation of all CDKs in eukaryotic cells requires phosphorylation at a conserved threonine (or serine) residue within their activation segment (T-loop) as well as cyclin binding. CDK-Activating Kinase (CAK) is an enzyme complex that is capable of phosphorylating CDKs at the T-loop and is essential for G1 and G2 CDK activities (Poon et al., 1993; Solomon et al., 1993; Fisher and Morgan 1994; Matsuoka et al., 1994). The CAK is composed of CDK7, cylin H, and Mat 1 (Nigg 1996). Cyclin H is a regulatory subunit of CDK7 and Mat 1 is an assembly factor for the complex (Fisher et al., 1995; Tassan et al., 1995; Devault et al., 1995) and regulates CAK substrate specificity (Yankulov and Bentley 1997).

The importance of CAK in ATRA-induced cell differentiation has been demonstrated in several studies. First, CAK is able to interact with and phosphorylate RAR *in vitro* (Rochette-Egly et al., 1997). Second, it has been observed that the mouse cells lacking Mat1 lost ability enter S phase and exhibited defects in phosphorylation of RNA polymerase II (Rossi et al., 2001). Subsequently, a decreased CAK pholation of RAR accompanied with differentiation has been observed in leukemic HL-60 cells treated with ATRA (Wang et al., 2002). In human, the formation of PML-RARα in APL patients blocks myeloid differentiation and suppresses cell apoptosis (Slack and Gallagher 1999; Melnick and Licht 1999, Jing Y et al., 2003). APL cells treated with ATRA exhibits the dissociation of PML/RARα from CAK. As a result of the dissociation, MAT1 is degraded and PML/RARα is converted from the hyperphosphorylation to the hypophosphorylation, which leads to G1 arrest and cell differentiation (He et al., 2004; Wang et al., 2006).

ATRA also regulates the expression of G1 CDK and CDK inhibitors. By using multiple approaches, Dimberg et al. demonstrated that ATRA-treated U937 cells showed G1/G0 arrest with downregulation of mRNA for cyclin E, D3, cyclin A, B and upregulation of p21 and p27 proteins (Dimberg et al., 2002). In another report, downgulation of cyclin E and cdk2 by ATRA in two AML cell lines, NB4 and HL-60, is a result of posttranslational modifications due to the activation of the ubiquitin-proteasome pathway (Fang et al. 2010).

It is well documented that as a transcriptional activator, c-Myc plays an important role in cell cycle progress and tumorigenesis and deregulation of c-Myc by ATRA has been observed in several cases (Xu et al., 2009; Imran et al. 2012). Since the downregulation of c-Myc precedes the G1/G0 arrest, it is possible that the inhibition of c-Myc by ATRA downregulates CDKs and upregulates CDK inhibitors (see above), which leads to the arrest of cell proliferation and induction of the cell differentiation. Since c-Myc can directly interact with RARα, causing either differentiation or proliferation, thus a distinct mechanism for RAR regulation by c-Myc has been reported, in which c-Myc-RARα complex either activates or inhibits differentiation target gene dependent on the c-Myc phosphorylation status. Unphosphorylated c-Myc-RARα represses the expression of RAR targets required for differentiation, thereby promoting cell proliferation, whereas phosphorylation of c-Myc by the kinase Pak2, the c-Myc/RARα complex activates transcription of those same genes to stimulate differentiation (Uribesalgo et al. 2012).

There is evidence that cell cycle regulatory molecules are also involved in vitamin D3-induced cell differentiation. For example, 1, 25(OH)$_2$D$_3$ induced monocytic differentiation of NB4 leukemia cells is accompanied by downregulation of CDK1, CDK2, CDK4, cyclin E and cycloin D3 (Clack et al., 2004), whereas the HL-60 myeloid leukemia cells treated with 1,25(OH)$_2$D$_3$ showed an increase in the level of CDK inhibitors (p21 and p27) and a decrease in the level of pRb phosphorylation. The dephophorylated pRb bound to transcription factor E2F1 activates cell differentiation pathway (Wang et al., 2009). The involvement of cell cycle regulatory molecules in the induction of the cell differentiation by vitamin D3 is not limited to the leukemia cells, because the changes in the expression of several CDKs and CDK inhibitors have also been observed in BxPC-3 human pancreatic carcinoma cells in response to vitamin D analog, MART-10 (Chiang et al., 2013).

INTRACELLULAR SIGNALING PATHWAYS

1. Raf-MEK-ERK

Activation of the mitogen activated protein kinase (MAPK) pathway following a ligand binding to various receptors at the cell surface has been correlated with numerous cellular responses, including proliferation, differentiation, and regulation of specific metabolic pathways in different cell types. Studies on fibroblasts and hematopoietic cells suggest that the activation of Raf, MEK, and ERK appears to be a linear pathway that is stimulated by growth factor receptors and leads to cell proliferation. The 44 kDa MAPK (ERK1) and 42 kDa MAPK (ERK2) are phosphorylated and activated by highly specific MEK1 and MEK2.The duration of MAPK activation is now known to determine whether a stimulus induces proliferation or differentiation (Seger and Krebs 1995; Marshall 1995). Landmark studies on PC 12 epithelial cells stimulated with growth factors revealed that cell differentiation requires prolonged activation of MAPK (lasting hours to days), whereas the transient activation of MAPK is, most likely, linked to cell proliferation (Traverse et al., 1992; Qiu and Green 1992; Yamada et al.,1996). Subsequently sustained activation of MEK-ERK signals associated with megakaryocytic differentiation in the K562 cell line (Whalen et al., 1997) and macrophage-like differentiation of TF-1a cells (Hu et al., 2000) were reported. In these studies, active MEK and ERK remained elevated for at least 2 h and returned to near-basal level by 24 h in response to PMA stimulation. However, megakaryocytic differentiation in the human erythroid/megakaryocytic cell line, HEL, appears to be dependent on protein kinase c (PKC) activation (Hong et al., 1996). Since PKC is capable of phosphorylating Raf (Sozeri et al., 1992), which is upstream of MEK, it is possible that PKC-induced differentiation may be through the activation of the MAPK pathway.

Activation of MEK and ERK by RA or ATRA has been well documented. In HL-60 human myeloblastic leukemia cells, RA activated ERK before it caused myeloid differentiation, hypophosphorylation of pRb, and cell cycle arrest. MEK inhibitor (PD98059) blocked RA-induced cell differentiation and growth arrest as well as RB hypophosphorylation (Yen et al., 1998). These data demonstrated that MEK-ERK2 activation is required for RA-induced differentiation.

The RA-induced ERK2 activation appears RA receptor-dependent, because a retinoid that selectively binds RAR-y, which is not expressed in HL-60 cells, had no effect on ERK2 activation. However, it is not clear whether the RA-induced MEK-ERK signal is RA-induced transcription dependent (Yen et al., 1998). RA also activates Raf, the upstream of MEK. Surprisingly, the phosphorylation of Raf (12-24 h after stating RA treatment) occurs much later than the phosphorylation of MEK (4 h after added RA) and parallels the differentiation process of the cells, suggesting that the Raf activation is MEK-dependent and is responsible for RA-induced differentiation. This signaling pathway is different from a typical growth factor-induced MAPK signaling cascade (Hong et al., 2001; Wang and Yen 2008).

Additional evidence for this non-typical pathway is that ATRA-induced phosphorylation of Raf1 was translocated into the nucleus during differentiation of HL-60 cells (Smith et al., 2009). Another player in MAPK-induced differentiation is Src. The Src or SFK (Src family kinase) is a family of non-receptor tyrosine kinases. It has been observed that hyperactivity of several members in this family is associated with acute and chronic myeloid leukemias (Dos Santos et al., 2008). By using SFK inhibitors, a recent study suggests that the persistent activation of MAPK induced by ATRA may also be regulated by SFK members (Congleton et al., 2012).

Although 1, 25(OH)$_2$D$_3$-induced differentiation and inhibition of human myeloid leukemia cells is primarily mediated through its nuclear receptor followed by target gene transcription, recent studies found that 1, 25(OH)$_2$D$_3$ can rapidly activate intracellular signaling, which is independent of its transcription activation (Zanello et al., 2004; Erben et al., 2002; Hughes et al., 2006).

The first report of activation of MAPK by 1, 25(OH)$_2$D$_3$, which is correlated with early HL-60 cell differentiation, was reported in 1997. In this report the activated MAPK was rapid and transient and found primarily in the nucleus (Marcinkowska et al., 1997; Wang and Studzinski 2001). This non-genomic effect of 1, 25(OH)$_2$D$_3$ requires presence of VDR but not RXR (Hughes and Brown 2006). Further studies suggest that activation of Raf, but not MEK or ERK, is required for 1,25 (OH)$_2$D$_3$-induced differentiation of HL60 cells (Wang and Studzinski 2006). In these studies MEK and ERK was only activated transiently upon 1,25(OK)$_2$D$_3$ treatment, whereas Raf-1 signal lasted at least for 96 hours, which was correlated with phenotypic changes of HL 60 cells. Furthermore, transfection of Raf-1 gene enhanced $1,25(OH)_2D_3$-induced differentiation and inhibition of Raf-1 blocked the cellular differentiation.

2. Protein Kinase (PKC) Isoforms

PKC is a member of the serine/threonine kinase superfamily. PKC phosphorylates a variety of target proteins and regulates a variety of cellular effects including cell proliferation, differentiation, apoptosis, aging, inflammation, and immunity. PKC itself is activated by diacylglycerol (DAG) and calcium ions (Ca^{2+}). Evidence of PKC-induced cell differentiation was originally obtained from numerous studies with phorbol-12-myristate-13-acetate (PMA)-, or 12-O-tetradecanoylphorbol 13-acetate (TPA)-induced leukemic cell differentiation (Hooper et al., 1989; Aihara et al., 1991; Slapak et al., 1993; Macfarlane et al., 1994; Hong et al., 1996). In these studies, persistent activation of PKC was detected in the leukemia cells treated with either PMA or TPA, and was correlated closely with cell differentiation and phenotypic changes. Additional support for the role of PKC in cell differentiation is the detection of the differences of the PKC activity in KG1 and KG1a cells (Hooper et al., 1989). The KG-1 cell line is composed predominantly of myeloblasts and promyelocytes (Koeffler and Golde 1978). KG-1a (Koeffler et al., 1980) is a subline isolated from KG-1. The KG-1myeloid leukemia cell line can be induced to differentiate into macrophages by TPA, while the KG-1a cell line is resistant to these differentiative effects of TPA. By performing *in vitro* PKC activity assay using type III-S histone as a substrate, Hooper and his colleagues found that KG-1a cells have a significant low level of PKC activity, suggesting the close correlation between the PKC level and cell differentiation (Hooper et al., 1989). Subsequently, $1,25(OH)_2D_3$-induced-PKC activation during cell differentiation has been observed by a number of groups (Obeid et al., 1990; Berry et al., 1996; Simpson et al., 1998). The persistent activation of PKC in the cells treated with 1,25(OH)2D3 occurred within 6 hours and lasted for several days. These observations are similar to those noted previously for PMA- and TPA-induced activation of PKC. In contrast, inhibition of PKC stops either $1,25(OH)_2D_3$ or ATRA-induced HL-60 myeloid differentiation (Kim et al., 2009) and K562 megakarocytic formation (Nakatake et al 2007). The role of PKC is also supported by the study on a variant of HL-60 cell line. This variant cell line is resistant to $1,25(OH)_2D_3$ and PKC in these cells is not able to translocate to the membrane from the cytosol. As a result, there is no differentiation to occur (Slapak et al., 1993). PKC has many isoforms and not all the isoforms have the same function in cell differentiation. One study suggests that it is PKCβ, rather than the PKCα, responsible for $1,25(OH)_2D_3$-induced differentiation, since pretreatment of the cells with antisense oligos against PKCβ, but not PKCα, aborted $1,25(OH)_2D_3$-induced differentiation (Simpson et al., 1998).

It has long been recognized that retinoids exert some of their effects on differentiation and growth inhibition of malignant cells partially due to their ability to activate PKC (Cho et al., 1997; Carter et al., 1998). In NB4 leukemia cells, ATRA and PMA, but not $1,25(OH)_2D_3$, activated PKCδ and a PKCδ-specific inhibitor Rottlerin almost removed the ATRA- and PMA-induced the cell differentiation. This biologic effect is mediated through the inhibition of the PKC downstream molecules Phospholipid Scramblase 1 (PLSCR1) (Zhao et al., 2004a). By performing gel shift assay and chromatin immunoprecipitation Kambhampati and his colleagues demonstrated that activated PKCδ is associated with RARα and bound to RAREs with other RA-inducible proteins. Pharmacological inhibition of PKCδ activity blocked the induction of cell differentiation and growth inhibition of NB-4 blast cells, indicating a critical role of PKCδ in ATRA–induced differentiation of acute myeloid leukemia cells (Kambhampati et al., 2003). Such a role is not limited to blood cells, because the activation of PKCδ has also been observed in ATRA-induced differentiation of neuron cells (Nitti et al., 2010). The Statins are a class of drugs widely used to manage hypercholesterolemia. These drugs function as competitive inhibitors of HMG-CoA (3-hydroxy-3-methyl-glutaryl-coenzyme A) reductase, the rate-limiting enzyme of cholesterol biosynthesis. Recently, it was reported that these drugs promote NB4 cell differentiation via activation of PKCδ (Sassano et al., 2012). The first evidence for the physical interaction of ATRA with PKC was reported in 2000, in which year a binding site for ATRA in PKCα was identified based on amino acid alignments and photo affinity labeling. However, instead of activation, the binding of ATRA to the PKCα decreases the PKC activity (Radominska-Pandya et al., 2000). Subsequently, it was shown that there are two binding sites for ATRA on the C2-domain of PKC (Ochoa et al., 2003).

3. Phosphatidylinositol 3-Kinase (PI3-K)-AKT Pathway

This pathway begins with receptor–induced activation of phosphatidylinositol 3-kinase (PI3-K). PI3-K then catalyzes the addition of a phosphate group to phosphatidylinositol-4, 5-bisphosphate (PIP2) to convert it to phosphatidylinol-3,4,5-trisphoshate (PIP3). PIP3 in turn activates AKT, also known as Protein Kinase B (PKB). The activation of AKT suppresses apoptosis and promotes cell growth. Several recent studies have found that the PI3K/AKT pathway is activated in AML cells (Min et al., 2003; Kubota et al., 2004; Xu et al., 2003; Zhao et al 2004b), suggesting that the overgrowth of the leukemia cells is linked to the high expression of PI3K/AKT pathway. However, the role of AKT in ATRA-induced differentiation is far from understood with stimulatory, inhibitory, or neutral effects having been reported.

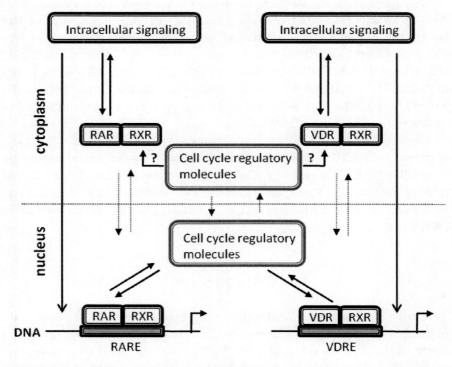

Figure 3. Diagram of the interactions between the ARTA and 1,25(OH)$_2$D$_3$ receptors and other molecules. → Signaling pathway → shuttling route.

These different effects appear to be linked to cell types tested. For example, it is well known that steroid sulphatase is one of hematopoietic cell differentiation makers. Upregulation of steroid sulphatase by ATRA and vitamin D3 has been observed in several human myeloid leukemia cell lines including HL60, U937 and THP1 (Hughes et al., 2001; 2005; Hughes and Brown 2006). By using pharmacological inhibitors, it was demonstrated that the increase of steroid sulphatase levels is correlated with the inhibition of the PIK-3-AKT pathway and NF-Kβ activity (Hughes et al., 2008.), suggesting that the PIK-3-AKR pathway suppresses ATRA activity. In contrast, The AKT activity was found to be required for ATRA-mediated expression of CD11b in leukemic NB4 (Matkovic et al., 2006) and neuron cell differentiation (Qiao et al., 2012). In an earlier study, ATRA induced phosphorylation of AKT in NB4 and primary APL cells, however, the cells pre-treated with PI3-K inhibitors had no effect on ATRA-induced differentiation (Billottet et al, 2009).

Although there are conflicting reports on the up-or down-regulation of AKT by vitamin D3, most likely, the PI3-K-AKT pathway has no direct effect on vitamin D3-induced differentiation of HL-60 cells. In one report, the phosphorylation and activity of AKT were found to be increased in HL-60 cells treated with $1,25(OH)_2D_3$. However, the activation of AKT is not linked to $1, 25(OH)_2D_3$ induced differentiation since a combination of $1,25(OH)_2D_3$ with PI3K inhibitor Ly294002 enhanced anti-proliferation and induction of differentiation (Zhang et al., 2006).

In another report, HL-60 cells treated with $1,25(OH)_2D_3$ had low expression of AKT, weakened Akt-Raf1 interaction, and enhanced expression of the Raf/MEK/ERK pathway. Only the activation of MEK and ERK was directly responsible for vitamin D3-induced differentiation, because the inhibition of MEK/ERK suppressed $1,25(OH)_2D_3$-induced differentiation (Wang et al., 2009).

CONCLUSION

Leukemic transformation results in an excess of immature cells with reduced ability for differentiation. Thus, differentiation induction has been a major strategy of cancer treatment. ATRA, the vitamin A derivative, and $1,25(OH)_2D_3$, the metabolite of vitamin D3, have been used as differentiation therapy in APL and non-APL leukemia cells over the past decades. The effects of ATRA and $1,25(OH)_2D_3$ are known to be mediated by their corresponding nuclear receptors (RARs and RXRs or VDR and RXRs) that belong to the type

II superfamily of nuclear hormone receptors. Upon ligand binding, these receptors interact with retinoic acid responsive elements (RARE) or vitamin D responsive elements (VDRE) and trigger the formation of coactivator complexes for the activation of transcription. A number of target genes by ATRA and $1,25(OH)_2D_3$ have been identified in recent years, which includes UBE1L, UBE2D3, C/EBPs, TRAIL, FOXOs, CAMPs, OLFM4, c-Myc, and cell cycle regulatory molecules (such as p21 and p27). The interaction between the nuclear receptors and cell cycle regulatory molecules may be transcription-dependent and -independent. For example, it has been demonstrated that c-Myc, p21, and p27 are target genes of ATRA and $1,25(OH)_2D_3$, whereas the phosphorylation or dephosphorylation of pRb, inactivation of CDKs or cylclins could be transcription independent. There is now emerging evidence indicating that intracellular signaling molecules are also involved in ATRA and 1,25(OH)2D3 induced differentiation of leukemia cells. There are several major intracellular signal transduction pathways activated by either ATRA or $1,25(OH)_2D_3$ such as Raf-MEK-ERK, PI3-K-AKT, and PKC pathways. The observation that ATRA and $1,25(OH)_2D_3$ nuclear receptors can be shuttled between the cytosol and the nucleus provides a rational basis for the interaction with intracellular signaling molecules. Figure 3 illustrates diagrammatically the interactions among the nuclear receptors, intracellular signaling pathways, and cell cycle regulatory molecules. Although we have accumulated much knowledge on the mechanisms responsible for ATRA and $1,25(OH)_2D_3$ induced differentiation of leukemia cells, a number of questions remain to be answered in future studies. For example, there are at least 160 ATRA targets genes and only a few of them have been well studied. The rest of the target genes need to be identified and additional target genes should be uncovered. The non-genomic pathways activated by ATRA and $1,2(OH)_2D_3$ have been intensively studied in recent years, however, the precise mechanisms for the interactions between the ATRA and $1,2(OH)_2D_3$ nuclear receptors and the intracellular signaling pathways described above need to be defined in more detail. Finally, it is unclear which gene(s) or signal molecule(s) play a critical role in determining the differentiation of AML cells into eithergranulocytes or monocytes/macrophages.

REFERENCES

Abe E, Miyaura C, Sakagami H, Takeda M, Konno K, Yamazaki T, Yoshiki S, Suda T (1981). Differentiation of mouse myeloid leukemia cells induced by 1-alpha,25-dihydroxyvitamin D3. *Proc. Natl. Acad. Sci. USA.* 78: 4990–4994.

Aihara H, Asaoka Y, Yoshida K, Nishizuka Y (1991). Sustained activation of protein kinase C is essential to HL-60 cell differentiation to macrophage. *Proc. Natl. Acad. Sci. USA.* 88:11062-11066.

Almasan A, Mangelsdorf DJ, Ong ES, Wahl GM, Evans RM. (1994). Chromosomal localization of the human retinoid X receptors. *Genomics.* 20:397-403.

Altucci L, Rossin A, Raffelsberger W, Reitmair A, Chomienne C, Gronemeyer H (2001). Retinoic acid-induced apoptosis in leukemia cells is mediated by paracrine action of tumor-selective death ligand TRAIL. *Nat. Med.* 7:680-686.

Berry DM, Antochi R, Bhatia M, Meckling-Gill KA. (1996).1,25-Dihydroxyvitamin D3 stimulates expression and translocation of protein kinase Calpha and Cdelta via a nongenomic mechanism and rapidly induces phosphorylation of a 33-kDa protein in acute promyelocytic NB4 cells. *J. Biol. Chem.* 271:16090-16096.

Billottet C, Banerjee L, Vanhaesebroeck B, Khwaja A. (2009). Inhibition of class I phosphoinositide 3-kinase activity impairs proliferation and triggers apoptosis in acute promyelocytic leukemia without affecting atra-induced differentiation. *Cancer Res.* 69:1027-1036.

Birkenkamp KU, Coffer PJ. (2003). Regulation of cell survival and proliferation by the FOXO (Forkheadbox, class O) subfamily of Forkhead transcriptionfactors. *Biochem. Soc. Trans.* 31(pt1):292-297.

Bobilev I, Novik V, Levi I, Shpilberg O, Levy J, Sharoni Y, Studzinski GP, Danilenko M. (2011). The Nrf2 transcription factor is a positive regulator of myeloid differentiation of acute myeloid leukemia cells. *Cancer Biol. Ther.*11:317-29.

Botling J, Oberg F, Törmä H, Tuohimaa P, Bläuer M, Nilsson K. (1996). Vitamin D3- and retinoic acid-induced monocytic differentiation: interactions between the endogenous vitamin D3 receptor, retinoic acid receptors, and retinoid X receptors in U-937 cells. *Cell Grow Differ.* 7:1239-1249.

Breitman TR, Collins SJ, Keene BR. (1981). Terminal differentiation of human promyelocytic leukemic cells in primary culture in response to retinoic acid. *Blood* 57:1000–1004.

Breitman TR, Selonick SE, Collins SJ. (1980). Induction of differentiation of the human promyelocytic leukemia cell line (HL-60) by retinoic acid. *Proc. Natl. Acad. Sci. USA.* 77:2936–2940.

Brown G, Bunce C, Rowlands D, Williams G. (1994). All-*trans* retinoic acid and 1α,25-di-hydroxyvitaminD3 co-operate to promote differentiation of the human promyeloid leukaemia cell line HL60 to monocytes. *Leukemia* 8:806–815

Brown TR Stonehouse TJ, Branch JS, Brickell PM, Katz DR. (1997). Stable transfection of U937 cells with sense or antisense RXR-alpha cDNA suggests a role for RXR-alpha in the control of monoblastic differentiation induced by retinoic acid and vitamin D. *Exp. Cell Res.* 236:94-102.

Brunet A, Bonni A, Zigmond MJ, Lin MZ, Juo P, Hu LS et al. (1999). Akt promotes cell survival by phosphorylating and inhibiting a Forkhead transcription factor. *Cell* 96: 857–868.

Capiati D, Rossi A, Picotti G, Benassati S, Boland R. (2004). Inhibition of serum-stimulated mitogen activated protein kinase by 1α,25(OH)$_2$-vitamin D$_3$ in MCF-7 breast cancer cells. *J. Cell Biochem.* 93: 334-397.

Carter CA, Parham GP, Chambers T. (1998). Cytoskeletal reorganization induced by retinoic acid treatment of human endometrial adenocarcinoma (RL95-2) cells is correlated with alterations in protein kinase C-alpha. *Pathobiology.* 66:284-92.

Chiang KC, Yeh CN, Hsu JT, Yeh TS, Jan YY, Wu CT, Chen HY, Jwo SC, Takano M, Kittaka A, Juang HH, Chen TC. (2013). Evaluation of the potential therapeutic role of a new generation of vitamin D analog, MART-10, in human pancreatic cancer cells in vitro and in vivo. *Cell Cycle.*12:1316-13125.

Cho Y, Tighe AP, Talmage DA. (1997). Retinoic acid induced growth arrest of human breast carcinoma cells requires protein kinase C alpha expression and activity. *J Cell Physiol.*172:306-313.

Clark CS, Konyer JE, Meckling KA. (2004). 1alpha,25-dihydroxyvitamin D3 and bryostatin-1 synergize to induce monocytic differentiation of NB4 acute promyelocytic leukemia cells by modulating cell cycle progression. *Exp. Cell Res.* 294:301-311.

Collins SJ, Robertson KA, Mueller L. (1990). Retinoic acid-induced granulocytic differentiation of HL-60 myeloid leukemia cells is mediated directly through the retinoic acid receptor (RAR-alpha). *Mol. Cell Biol.* 10:2154-63.

Collins SJ, Ruscetti FW, Gallagher RE, Gallo RC. (1978). Terminal differentiation of human promyelocytic leukemia cells induced by dimethyl sulfoxide and other polar compounds. *Proc. Natl. Acad. Sci. USA* 75:2458-2462.

Congleton C, MacDonald R, and Yen A (2012). Src Inhibitors, PP2 and Dasatinib, Increase Retinoic Acid-Induced Association of Lyn and c-Raf (S259) and Enhance MAPK Dependent Differentiation of Myeloid Leukemia Cells. *Leukemia:* 26: 1180-1188.

Devault A, Martinez AM, Fesquet D, Labbé JC, Morin N, Tassan JP, Nigg EA, Cavadore JC, Dorée M. (1995). MAT1 ('menage á trois') a new RING finger protein subunit stabilizing cyclin H-cdk7 complexes in starfish and Xenopus CAK. *EMBO J.* 14: 5027–5036.

de Thé H, Lavau C, Marchio A, Chomienne C, Degos L, Dejean A. (1991). The PML-RARa fusion mRNA generated by the t(15; 17) translocation in acute promyelocytic leukemia encodes a functionally altered RAR. *Cell* 66:675-684.

Dhandapani L, Yue P, Ramalingam SS, Sun SY. (2001). Retinoic acid enhances TRAIL-induced apoptosis in cancer cells by upregulating TRAIL receptor 1 expression. *Cancer Research* 71: 5245-5254.

Dhawan P, Wieder R, Christakos S. (2009). CCAAT enhancer-binding protein alpha is a molecular target of 1,25-dihydroxyvitamin D3 in MCF-7 breast cancer cells. *J. Biol. Chem.* 284:3086-3095.

Dijkers PF, Medema RH, Lammers JW, Koenderman L, Coffer PJ. (2000). Expression of the pro-apoptotic Bcl-2 family member Bim is regulated by the forkhead transcription factor FKHR-L1. *Curr. Biol.*10: 1201–1204.

Dimberg A, Bahram F, Karlberg I, Larsson L, Niscson K, Oberg F. (2002). Retinoic acid-induced cell cycle arrest of human myeloid cell lines is associated with sequential downb-regulation of c-Mycandcyclin E and posttranscriptional up-regulation of p27kip1. *Blood* 99: 2199-2206.

Dos Santos C, Demur C, Bardet V, Prade-Houdellier N, Payrastre B, Récher C. (2008). A critical role for Lyn in acute myeloid leukemia. *Blood* 111:2269-79.

Duprez E, Wagner K, Koch H, Tenen DG. (2003). C/EBPbeta: a major PML-RARA-responsive gene in retinoic acid-induced differentiation of APL cells. *EMBO J.* 22:5806-5816.

Erben RG, Soegiarto DW, Weber K, Zeitz U, Lieberherr M, Gniadecki R, Möller G, Adamski J, Balling R. (2002). Deletion of deoxyribonucleic acid binding domain of the vitamin D receptor abrogates genomic and nongenomic functions of vitamin D. *Mol. Endocrinol.* 16:1524-1537.

Fang Y, Zhou X, Lin M, Zhong L, Ying M, Luo P, Yang B, He Q. (2010). The ubiquitin-proteasome pathway plays an essential role in ATRA-induced leukemia cells G0/G1 phase arrest and transition into granulocytic differentiation. *Cancer Biology & Therapy* 10: 1157-1167.

Fisher, R.P., P. Jin, H.M. Chamberlin, and D.O. Morgan. (1995). Alternative mechanisms of CAK assembly require an assembly factor or an activating kinase. *Cell* 83: 47–57.

Fisher, R.P. and D.O. Morgan. (1994). A novel form of cyclin associates with MO15/CDK7 to from the CDK-activating kinase. *Cell* 78: 713–724.

Friedman AD. (2007). Transcriptional control of granulocyte and monocyte development. 26:6816-6828.

Fritsche J, Stonehouse TJ, Katz DR, Andreesen R, Kreutz M. (2000). Expression of retinoid receptors during human monocyte differentiation in vitro. *Biochem. Biophys. Res. Commun*.270:17-22.

Fujiki A, Imamura T, Sakamoto K, Kawashima S, Yoshida H, Hirashima Y, Miyachi M, Yagyu S, Nakatani T, Sugita K, Hosoi H. (2012). All-trans retinoic acid combined with 5-Aza-2'-deoxycitidine induces C/EBPα expression and growth inhibition in MLL-AF9-positive leukemic cells. *Biochem. Biophys. Res. Commun.* 428:216-223.

Gocek E, Marchwicka A, Baurska H, Chrobak A, Marcinkowska E. (2012). Opposite regulation of vitamin D receptor by ATRA in AML cells susceptible and resistant to vitamin D-induced differentiation. *J. Steroid Biochem. Mol. Biol.* 132:220-226.

Gombart AF, Borregaard, N and Koeffler HP (2005). Human cathelicidin antimicrobial peptide (CAMP) geneis a direct target of the vitamin D receptor and isstrongly up-regulated in myeloid cells by 1,25-dihydroxyvitamin D3. *The FASEB J.* 119: 1067-1077.

Hackanson B, Bennett KL, Brena RM, Jiang J, Claus R, Chen SS, Blagitko-Dorfs N, Maharry K, Whitman SP, Schmittgen TD, Lübbert M, Marcucci G, Bloomfield CD, Plass C (2008). Epigenetic modification of CCAAT/enhancer binding protein alpha expression in acute myeloid leukemia. *Cancer Res.* 68:3142-3151

Hattori H, Zhang X, Jia Y, Subramanian KK, Jo H, Loison F, Newburger PE, Luo HR. (2007). RNAi screen identifies UBE2D3 as a mediator of all-trans retinoic acid-induced cell growth arrest in human acute promyelocytic NB4 cells. *Blood* 110: 640-650.

He Q, Peng H, Collines SJ, Triche TJ and Wu L. (2004). Retinoid-modulated MAT1 ubiquitination and CAK activity. *The FASEB J.* 18: 1734-1736.

Heyman RA, Mangelsdorf DJ, Dyck JA, Stein RB, Eichele G, Evans RM, Thaller C (1992). 9-cis retinoic acid is a high affinity ligand for the retinoid X receptor. *Cell.* 68:397-406.

Hmama Z, Nandan S, Sly L, Knutson K, Herrera-Velit P, Reiner N. (1999). 1α,25-dihydroxyvitamin D3-induced myeloid cell differentiation is regulated by a vitamin D receptor-phosphatidylinositol 3-kinase signalling complex. *J. Exp. Med.* 190:1583–1594.

Hong HY, Varvayanis S, Yen A (2001). Retinoic acid causes MEK-dependent RAF phosphorylation throughRARaplus RXR activation in HL-60 cells. *Differentiation* 68:55–66.

Hong Y, Martin JF, Vainchenker W, and Erusalimsky JD. (1996). Inhibition of protein kinase C suppresses megakaryocytic differentiation and stimulates erythroid differentiation in HEL cells. *Blood* 87: 123–131.

Honma Y, Hozumi M, Abe E, Konno K, Fukushima M, Hata S, Nishii Y, DeLuca HF, Suda T. (1983). 1-Alpha,25-dihydroxyvitamin D3 and 1-alpha-hydroxyvitamin D3 prolong survival time of mice inoculated with myeloid leukemia cells. *Proc. Natl. Acad. Sci. USA.* . 80: 201–204.

Hooper WC, Abraham RT, Ashendel CL, Woloschak GE. (1989). Differential responsiveness to phorbol esters correlates with differential expression of protein kinase C in KG-1 and KG-1a human myeloid leukemia cells. *Biochim. Biophys. Acta.* 1013:47-54.

Hu X, Moscinski LC, Valkov NI, Fisher AB, Hill BJ and Zuckerman KS. (2000) Prolonged Activation of the Mitogen-activated Protein Kinase Pathway Is Required for Macrophage-like Differentiation of a Human Myeloid Leukemic Cell Line. *Cell Growth & Differentiation* 11: 191-200.

Huang ME, Ye YC, Chen SR, Zhao JC, Gu LJ, Cai JR, Zhao L, Xie JX, Shen ZX, Wang ZY. (1987). All-trans retinoic acid with or without low dose cytosine arabinoside in acute promyelocytic leukemia. Report of 6 cases. *Chin. Med. J.* (Engl) 100:949–953.

Huang ME, Ye YC, Chen SR, Chai JR, Lu JX, Zhoa L, Gu LJ, Wang ZY. (1988). Use of all-trans retinoic acid in the treatment of acute promyelocytic leukemia. *Blood* 72:567–572.

Hughes PJ, and Brown G. (2006). 1Alpha,25-dihydroxyvitamin D3-mediated stimulation of steroid sulphatase activity in myeloid leukaemic cell lines requires VDRnuc-mediated activation of the RAS/RAF/ERK-MAP kinase signalling pathway. *J. Cell Biochem.* 98:590-617.

Hughes PJ, Lee JS, Reiner NE, Brown G. (2008). The vitamin D receptor-mediated activation of phosphatidylinositol 3-kinase (PI3Kalpha) plays a role in the 1alpha,25-dihydroxyvitamin D3-stimulated increase in steroid sulphatase activity in myeloid leukaemic cell lines. *J. Cell Biochem.* 103:1551-1572.

Hughes PJ, Steinmeyer A, Chandraratna RA, Brown G. (2005). 1alpha,25-dihydroxyvitamin D3 stimulates steroid sulphatase activity in HL60 and

NB4 acute myeloid leukaemia cell lines by different receptor-mediated mechanisms. *J. Cell Biochem.* 94:1175-1189.

Hughes PJ, Twist LE, Durham J, Choudhry MA, Drayson M, Chandraratna R, Michell RH, Kirk CJ, Brown G. (2001). Up-regulation of steroid sulphatase activity in HL60 promyelocytic cells by retinoids and 1alpha,25-dihydroxyvitamin D3. *Biochem. J.* 355(Pt 2):361-371.

Idres N, Benoît G, Flexor MA, Lanotte M, Chabot GG. (2001). Granulocytic differentiation of human NB4 promyelocytic leukemia cells induced by all-trans retinoic acid metabolites.*Cancer Res.*61:700-705.

Imran M, Park TJ, Lim IK. (2012). TIS21/BTG2/PC3 enhances downregulation of c-Myc during differentiation of HL-60 cells by activating Erk1/2 and inhibiting Akt in response to all-trans-retinoic acid. *Eur. J. Cancer.* 48:2474-2485.

Jackman M, Kubota Y, den Elzen N, Hagting, A and Pines J (2002). Cyclin A- and Cyclin E-Cdk Complexes Shuttle between the Nucleus and the Cytoplasm. *Mol. Bio. Cell* 13: 1030–1045.

Jamshidi F, Zhang J, Harrison JS, Wang X, Studzinski GP. (2008). Induction of differentiation of human leukemia cells by combinations of COX inhibitors and 1,25-dihydroxyvitamin D3 involves Raf1 but not Erk 1/2 signaling. *Cell Cycle.* 7:917-924.

Ji Y, Studzinski GP. (2004). Retinoblastoma protein and CCAAT/enhancer-binding protein beta are required for 1,25-dihydroxyvitamin D3-induced monocytic differentiation of HL60 cells. *Cancer Res.* 64:370-737.

Jing Y, Xia L, Lu M and Waxman S (2003). The cleavage product DPML–RARa contributes to all-trans retinoic acid-mediated differentiation in acute promyelocytic leukemia cells. *Oncogene* 22: 4083-4091.

Kakizuka A, Miller WH, Umesono K, Warrell RP, Frankel SR, Murty VVVS, Dmitrovsky E, Evans RM. (1991). Chromosomal translocation t(15; 17) in human acute promyelocytic leukemia fuses RARa with a novel putative transcription factor, PML. *Cell* 66:663-674.

Kambhampati S, Li Y, Verma A, Sassano A, Majchrzak B, Deb DK, Parmar S, Giafis N, Kalvakolanu_DV, Rahman A,Uddin S, Minucci S, Tallman MS, Fish EN,and Platanias LC. (2003). Activation of Protein Kinase C by All-*trans*-retinoic Acid. *J. Biol. Chem* 278: 32544–32551.

Kandilci A, Grosveld GC (2009). Reintroduction of CEBPA in MN1-overexpressing hematopoietic cells prevents their hyperproliferation and restores myeloid differentiation. *Blood* 114:1596-1606.

Kastner P, Perez A, Lutz Y, Rochette-Egly C, Gaub MP, Durand B, Lanotte M, Berger R, Chambon P (1992). Structure, localization and transcriptional properties of two classes of retinoic acid receptor αfusion proteins in acute promyelocytic leukemia (APL): Structural similarities with a new family of oncoproteins. *EMBO J.* 11:629-642.

Kim SH, Cho SS, Simkhada JR, Lee HJ, Kim SW, Kim TS, Yoo JC. (2009). Enhancement of 1,25-dihydroxyvitamin D3- and all-trans retinoic acid-induced HL-60 leukemia cell differentiation by Panax ginseng. *Biosci. Biotechnol. Biochem.* 73:1048-1053.

Kim Y, MacDonald P, Dedhar S, Hruska K. (1996). Association of 1α,25(OH)-dihydroxyvitamin D3-occupied vitamin D receptors with cellular membrane acceptance sites. *Endocrinology* 137: 3649-3658.

Kitareewan S, Pitha-Rowe I, Sekula D, Lowrey CH, Nemeth MJ, Golub TR, Freemantle SJ, Dmitrovsky E. (2002). UBE1L is a retinoid target that triggers PML/RARalpha degradation and apoptosis in acute promyelocytic leukemia. *Proc. Natl. Acad. Sci. USA* 99:3806-3811.

Koeffler HP, Billing R, Lusis AJ, Sparkes R and Golde DW. (1980). An undifferentiated variant derived from the human acute myelogenous leukemia cell line (KG-1). *Blood* 56: 265-273.

Koeffler and Golde (1978). Acute myelogenous leukemia: a human cell line responsive to colony-stimulating activity. *Science* 200: 1153-1154.

Kok K, Hofstra R, Pilz A, van den Berg A, Terpstra P, Buys CH, Carritt B. (1993). A gene in the chromosomal region 3p21 with greatly reduced expression in lung cancer is similar to the gene for ubiquitin-activating enzyme. *Proc. Natl. Acad. Sci. USA.* 90:6071-6075.

Komatsu N, Watanabe T, Uchida M, Mori M, Kirito K, Kikuchi S, Liu Q, Tauchi T, Miyazawa K, Endo H, Nagai T, Ozawa K. (2003). A member of Forkhead transcription factor FKHRL1 is a downstream effector of STI571-induced cell cycle arrest in BCR-ABL-expressing cells. *J. Biol. Chem.* 278:6411-6419.

Kops GJ, Medema RH, Glassford J, Essers MA, Dijkers PF, Coffer PJ et al. (2002) Control of cell cycle exit and entry by protein kinase B-regulated forkhead transcription factors. *Mol. Cell. Biol.* 22, 2025–2036.

Krauss G. (2001). Biochemistry of Signal transduction and regulation. Second Edition (Wiley-VCH), pp 167-168.

Kubota Y, Ohnishi H, Kitanaka A, Ishida T, Tanaka T. (2004). Constitutive activation of PI3K is involved in the spontaneous proliferation of primary acute myeloid leukemia cells: direct evidence of PI3K activation. *Leukemia.* 18:1438-1440.

Lee YK, Choi YH, Chua S, Park YJ, Moore DD. (2006). Phosphorylation of the hinge domain of the nuclear hormone receptor LRH-1 stimulates transactivation. *J. Biol. Chem.* 281:7850-7855.

Levin AA, Sturzenbecker LJ, Kazmer S, Bosakowski T, Huselton C, Allenby G, Speck J, Kratzeisen C, Rosenberger M, Lovey A, et al (1992). 9-cis retinoic acid stereoisomer binds and activates the nuclear receptor RXR alpha. *Nature* 355:359-361.

Liu TX, Zhang JW, Tao J, Zhang RB, Zhang QH, Zhao CJ, Tong JH, Lanotte M, Waxman S, Chen SJ, Mao M, Hu GX, Zhu L, Chen Z. (2000). Gene expression networks underlying retinoic acid-induced differentiation of acute promyelocytic leukemia cells. *Blood* 96:1496-1504.

Liu W, Lee H , Liu, Y, Wang R Rodgers GP (2010). Olfactomedin 4 is a novel target gene of retinoic acids and 5-aza-2'-deoxycytidine involved in human myeloid leukemia cell growth, differentiation, and apoptosis. *Blood* 116: 4938-4947.

Macfarlane DE, Manzel L (1994) Activation of beta-isozyme of protein kinase C (PKC beta) is necessary and sufficient for phorbol ester-induced differentiation of HL-60 promyelocytes. Studies with PKC beta-defective PET mutant. *J. Biol. Chem.* 269:4327-4331.

Mangelsdorf D, Koeffler HP, Donaldson C, Pike J, Haussler M. (1984). 1,25-dihydroxyvitamin D3-induced differentiation in a human promyelocyticleukaemia cell line (HL-60): receptor-mediated maturation to macrophage-like cells. *J. Cell Biol* 98:391–398.

Mangelsdorf DJ, Thummel C, Beato M, Herrlich P, Schütz G, Umesono K, Blumberg B, Kastner P, Mark M, Chambon P, Evans RM. (1995). The nuclear receptor superfamily: the second decade. *Cell* 83:835-839.

Marcinkowska E, Garay E, Gocek E, Chrobak A, Wang X, Studzinski GP. (2006). Regulation of C/EBPbeta isoforms by MAPK pathways in HL60 cells induced to differentiate by 1,25-dihydroxyvitamin D3. *Exp. Cell Res.* 312:2054-2065

Marcinkowska E, Wiedłocha A, Radzikowski C. (1997). 1,25-Dihydroxyvitamin D3 induced activation and subsequent nuclear translocation of MAPK is upstream regulated by PKC in HL-60 cells. *Biochem. Biophys. Res. Commun.* 241:419-426.

Marshall, C. J. (1995). Specificity of receptor tyrosine kinase signaling: transient *versus* sustained extracellular signal-regulated kinase activation. *Cell* 80: 179–185.

Matkovic K, Brugnoli F, Bertagnolo V, Banfic H, Visnjic D. (2006) The role of the nuclear Akt activation and Akt inhibitors in all-trans-retinoic acid-differentiated HL-60 cells. *Leukemia.* 20:941-951.

Matsuoka, M., J.-Y.Kato, R.P. Fisher, D.O. Morgan, and C.S. Sherr. (1994). Activation of cyclin-dependent kinsase 4 (cdk4) by mouse MO15-associated kinase. *Mol. Cell Biol.*14: 7265–7275.

Mattei MG, Rivière M, Krust A, Ingvarsson S, Vennström B, Islam MQ, Levan G, Kautner P, Zelent A, Chambon P, et al. (1991). Chromosomal assignment of retinoic acid receptor (RAR) genes in the human, mouse, and rat genomes. *Genomics.* 10:1061-1069.

Medema RH, Kops GJ, Bos JL, Burgering BM. (2000). AFX-like Forkhead transcription factors mediate cell-cycle regulation by Ras and PKB through p27kip1. *Nature* 404: 782–787.

Melnick A, Licht JD. (1999). Deconstructing a disease: RARalpha, its fusion partners, and their roles in the pathogenesis of acute promyelocytic leukemia. *Blood* 93:3167-215.

Min YH, Eom JI, Cheong JW, Maeng HO, Kim JY, Jeung HK, Lee ST, Lee MH, Hahn JS, Ko YW. (2003). Constitutive phosphorylation of Akt/PKB protein in acute myeloid leukemia: its significance as a prognostic variable. *Leukemia.*17:995-997.

Miyamoto K, Araki KY, Naka K, Arai F, Takubo K, Yamazaki S, Matsuoka S, Miyamoto T, Ito K, Ohmura M, Chen C, Hosokawa K, Nakauchi H, Nakayama K, Nakayama KI, Harada M, Motoyama N, Suda T, Hirao A. (2007). Foxo3a is essential for maintenance of the hematopoietic stem cell pool. *Cell Stem. Cell* 1:101-102.

Modur V, Nagarajan R, Evers BM, Milbrandt J. (2002). FOXO proteins regulate tumor necrosis factor-related apoptosis inducing ligand expression. Implications for PTEN mutation in prostate cancer. *J. Biol. Chem.* 277: 47928–47937.

Nagamura-Inoue T, Tamura T, OzatoK. (2001). Transcription factors that regulate growth and differentiation of myeloid cells. *Int. Rev. Immunol.* 20:83-105

Nakamura N, Ramaswamy S, Vazquez F, Signoretti S, Loda M, Sellers WR. (2000). Forkhead transcription factors are critical effectors of cell death and cell cycle arrest downstream of PTEN. *Mol. Cell Biol.* 20: 8969–8982.

Nakatake M, Kakiuchi Y, Sasaki N, Murakami-Murofushi K, Yamada O. (2007). STAT3 and PKC differentially regulate telomerase activity during megakaryocytic differentiation of K562 cells. *Cell Cycle.* 6:1496-1501.

Nigg EA. (1996) Cyclin-dependent kinase-7: at the cross-roads of transcription, DNA-repair and cell-cycle control. *Curr. Opin Cell Biol.* 8: 312–317.

Nilsson B. (1984). Probable in vivo induction of differentiation by retinoic acid of promyelocytes in acute promyelocyticleukaemia. *Br. J. Haematol.* 57:365–371.

Nitti M, Furfaro AL, Cevasco C, Traverso N, Marinari UM, Pronzato MA, Domenicotti C (2010). PKC delta and NADPH oxidase in retinoic acid-induced neuroblastoma cell differentiation *Cell Signal.* 22:828-835.

Nizet V, Ohtake T, Lauth X, Trowbridge J, Rudisill J, Dorschner RA, Pestonjamasp V, Piraino J, Huttner K, Gallo RL (2001). "Innate antimicrobial peptide protects the skin from invasive bacterial infection". *Nature* 414: 454–457.

Nurse P. (2000). A long twentieth century of the cell cycle and beyond. *Cell* 100:71-78.

Obeid LM, Okazaki T, Karolak LA, and Hannun YA. (1990). Transcriptional regulation of protein kinase C by 1,25-Dihydroxyvitamin D3 in HL-60 cells. *J. Bio. Chem.* 265: 2370-2374.

Ochoa WF, Torrecillas A, Fita I, Verdaguer N, Corbalan-Garcia S, Gomez-Fernandez JC (2003). Retinoic acid binds to the C2-domain of protein kinase Cα. *Biochemistry* 42: 8774-8779.

Ohata M, Yamauchi M, Takeda K, Toda G, Kamimura S, Motomura K, Xiong S, Tsukamoto H. (2000). RAR and RXR expression by Kupffer cells. *Exp. Mol. Pathol.* 68:13-20.

Pandolfi PP, Alcalay M, Mencarelli A, Biondi A, Lo Coco F, Grignani F, Pelicci PG. (1991):.Structure and origin of the acute promyelocytic leukemia myURARAcDNA and characterization of its retinoidbinding and transactivation properties. *Oncogene* 6: 1285-.

Park DJ, Chumakov AM, Vuong PT, Chih DY, Gombart AF, Miller WH Jr, Koeffler HP (1999). CCAAT/enhancer binding protein epsilon is a potential retinoid target gene in acute promyelocytic leukemia treatment. *J. Clin. Invest.* 103:1399-1408.

Pitha-Rowe I, Petty WJ, Kitareewan S, and Dmitrovsky E. (2003), Retinoid target genes in acute promyelocytic leukemia. *Leukemia* 17: 1723-1730.

Poon, R.Y.C., K. Yamashita, J.P. Adamczewszi, T. Hunt, and J. Shuttleworth. 1993. The cdc2-related protein p40MO15 is the catalytic subunit of a protein kinase that can activate p33cdk2 and p34cdc2. *EMBO J.* 12: 3123–3132.

Prüfer K, Barsony J. (2002). Retinoid X receptor dominates the nuclear import and export of the unliganded vitamin D receptor. *Mol Endocrinol.* 16:1738-1751.

Qiao J, Paul P, Lee S, Qiao L, Josifi E, Tiao JR, Chung DH (2012). PI3K/AKT and ERK regulate retinoic acid-induced neuroblastoma cellular differentiation. *Biochem. Biophys. Res. Commun.* 424:421-426.

Qiu, M, Green, S. H. (1992). PC12 cell neuronal differentiation is associated with prolonged p21ras activity and consequent prolonged ERK activity. *Neuron* 9:705–717.

Rachez C, Lemon BD, Suldan Z, Bromleigh V, Gamble M, Näär AM, Erdjument-Bromage H, Tempst P, Freedman LP (1999). Ligand-dependent transcription activation by nuclear receptors requires the DRIP complex. *Nature.* 398:824-828.

Racz A, Barsony J. (1999). Hormone-dependent translocation of vitamin D receptors is linked to transactivation. *J. Biol. Chem.* 274:19352-19360.

Radominska-Pandya A, Chen G, Czernik PJ, Little JM, Samokyszyn VM, Carteri CA, and Nowak G. (2000). Direct Interaction of All-*trans*-retinoic Acid with Protein Kinase C (PKC) *J. Biol. Chem.* 275: 22324–22330.

Rhinn M and Dolle P. (2012). Retinoic acid signaling during development. *Development* 139: 843-858, 2012.

Ricote M, Snyder CS, Leung HY, Chen J, Chien KR, Glass CK. (2006). Normal hematopoiesis after conditional targeting of RXRalpha in murine hematopoietic stem/progenitor cells. *J. Leukoc. Biol.* 80:850-861.

Roberts JM. (1999). Evolving ideas about cyclins. *Cell* 97:129-132.

Robertson KA, Emami B, Collins SJ. (1992). Retinoic acid-resistant HL-60R cells harbor a point mutation in the retinoic acid receptor ligand-binding domain that confers dominant negative activity. *Blood.* 80:1885-1889.

Robinson-Rechavi M, Escriva Garcia H, Laudet V. (2003). The nuclear receptor superfamily. *J. Cell Sci.* 116:585-586.

Rochette-Egly, C., Adam, S., Rossignol, M., Egly, J. M., and Chambon, P. (1997). Stimulation of RAR alpha activation function AF-1 through binding to the general transcription factor TFIIH and phosphorylation by CDK7. *Cell* 90: 97–107.

Rossi DJ, Londesborough A, Korsisaari N, Pihlak A, Lehtonen E, Henkemeyer M, Mäkelä TP. (2001). Inability to enter S phase and defective RNA polymerase II CTD phosphorylation in mice lacking Mat1. *EMBO J.* 20:2844-2856.

Rovera G, O'Brien TG, Diamond L (1979): Induction of differentiation in human promyelocytic leukemia cells by tumor promoters. *Science* 204:868-870.

SakoeY, Sakoe K, Kirito K,3 Ozawa K and Komatsu N. (2010). FOXO3Aas a key molecule for all-*trans* retinoic acid–induced granulocytic differentiation and apoptosis in acute promyelocytic leukemia. *Blood* 115: 3787-3795.

Salomoni P, Pandolfi PP. (2002).The role of PML in tumor suppression. *Cell*.108:165-170.

Sassano A, Altman JK, Gordon LI, Platanias LC. (2012). Statin-dependent activation of protein kinase Cδ in acute promyelocytic leukemia cells and induction of leukemic cell differentiation. *Leuk Lymphoma*. 53:1779-84.

Seger R and Krebs E G. (1995). The MAPK signaling cascade. *FASEB J.* 9: 726–735.

Simpson RU, O'Connell TD, Pan Q, Newhouse J, Somerman MJ (1998). Antisense oligonucleotides targeted against protein kinase Cbeta and CbetaII block 1,25-(OH)2D3-induced differentiation. *J. Biol. Chem.* 273:19587-19591.

Slack JL, Gallagher RE. (1999). The molecular biology of acute promyelocytic leukemia. *Cancer Treat Res.* 99:75-124.

Slapak CA, Kharbanda S, Saleem A, Kufe DW (1993). Defective translocation of protein kinase C in multidrug-resistant HL-60 cells confers a reversible loss of phorbol ester-induced monocytic differentiation. *J. Biol. Chem.* 268:12267-12273.

Smith J, Bunaciu RP, Reiterer G, Coder D, George T, Asaly M, and Yen A. (2009). Retinoic acid induces nuclear accumulation of Raf1 during differentiation of HL-60 cells. *Exp. Cell Res*. 315: 2241–2248.

Solomon MJ, Harper JW, and Shuttleworth J. (1993). CAK, the p34cdc2 activating kinase, contains a protein identical or closely related to p40MO15. *EMBO J*. 12: 3133–3142.

Sozeri, O., Vollmer, K., Liyanage, M., Frith, D., Kour, G., Mark, G. E.,andStabel, S. (1992) Activation of the c-raf protein kinase by protein kinase C phosphorylation. *Oncogene,* 7: 2259–2262.

Tanaka Y, Shima M, Yamaoka K, Okada S, Seino Y. (1992). Synergistic effect of 1,25-dihydroxyvitaminD3 and retinoic acid in inducing U937 cell differentiation. *J. Nutr. Sci. Vitaminol.* (Tokyo) 38:415–426.

Tassan, J.-P., M. Jaquenoud, A.M. Fry, S. Frutiger, G.J. Hughes, and E.A. Nigg. (1995). In vitro assembly of a functional human CDK7-cyclin H complex requires MAT1, a novel 36 kDa RING finger protein. *EMBO J.* 14: 5608–5617.

Tata JR. (2002). Signalling through nuclear receptors. *Nat. Rev. Mol. Cell Biol.*3:702-710.

Tothova Z, Kollipara R, Huntly BJ, Lee BH, Castrillon DH, Cullen DE, McDowell EP, Lazo-Kallanian S, Williams IR, Sears C, Armstrong SA, Passegué E, DePinho RA, Gilliland DG. (2007). FoxOs are critical mediators of hematopoietic stem cell resistance to physiologic oxidative stress. *Cell* 128:325-339.

Traverse S, Gomez, N., Paterson, H., Marshall, C., and Cohen, P. (1992). Sustained activation of the mitogen-activated protein (MAP) kinase cascademay be required for differentiation of PC12 cells. *Biochem. J., 288:*351–355.

Truong BT, Lee YJ, Lodie TA, Park DJ, Perrotti D, Watanabe N, Koeffler HP, Nakajima H, Tenen DG, Kogan SC. (2003). CCAAT/Enhancer binding proteins repress the leukemic phenotype of acute myeloid leukemia. *Blood* 101:1141-1148.

Tuteja G, Kaestner KH (2007a) SnapShot: forkhead transcription factors I. *Cell* 130, 1160.

Tuteja G, Kaestner KH (2007b) SnapShot: forkhead transcription factors II. *Cell* 131, 192.

Uribesalgo I, Benitah SA, Di Croce L. (2012). From oncogene to tumor suppressor: the dual role of Myc in leukemia. *Cell Cycle.* 11:1757-64.

Wang J Barsky LW, Shum CH, Jong A, Weinberg§ KI, Collins SJ, Triche TJ, and Wu L (2002). Retinoid-induced G1 arrest and differentiation activation areassociated with a switch to cyclin-dependent kinase-activating kinase hypophosphorylation of retinoic acid receptor. *J. Biol. Chem.* 277: 43369–43376.

Wang JG, Barsky LW, Davicioni E, Weinberg KI, Triche TJ, Zhang XK, Wu L (2006). Retinoic acid induces leukemia cell G1 arrest and transition into differentiation by inhibiting cyclin-dependent kinase-activating kinase binding and phosphorylation of PML/RARalpha. *FASEB J.* 20:2142-2144.

Wang J, Yen A . (2008). A MAPK-positive feedback mechanism for BLR1 signaling propels retinoic acid-triggered differentiation and cell cycle arrest. *J. Biol. Chem.* 283:4375-4386.

Wang J, Zhao Y, Kauss MA, Spindel S, Lian H. (2009). Akt regulates vitamin D3-induced leukemia cell functional differentiation via Raf/MEK/ERK MAPK signaling. *Eur. J. Cell Biol.* 88:103-15.

Wang X and Studzinski GP. (2001). Activation of extracellular signal-regulated kinases (ERKs) defines the first phase of 1,25-dihydroxyvitamin D3-induced differentiation of HL60 cells. *J. Cell Biochem.* 80: 471-482.

Wang X and Studzinski GP (2006). Raf-1 signaling is required for the later stages of 1,25-dihydroxyvitamin D3-induced differentiation of HL60 cells but is not mediated by the MEK/ERK module. *J. Cell Physiol.* 209:253-260.

Whalen, A. M., Galasinski, S. G., Shapiro, P., Nahreini, T. S., and Ahn, N. G. (1997). Megakaryocytic differentiation induced by constitutive activation of mitogen-activated protein kinase kinase. *Mol. Cell. Biol.,* 17: 1947–1958.

Woods YL, Rena G. (2002). Effect of multiple phosphorylation events on the transcription factors FKHR, FKHRL1 and AFX. *Biochem. Soc. Trans.* 30:391-397.

Xu B, Liu P, Li J, Lu H. (2009). All-trans retinoic acid induces Thrombospondin-1 expression in acute promyelocytic leukemia cells though down-regulation of its transcription repressor, c-Myc oncoprotein. *Biochem. Biophys. Res. Commun.* 382:790-794.

Xu Q, Simpson SE, Scialla TJ, Bagg A, Carroll M. (2003). Survival of acute myeloid leukemia cells requires PI3 kinase activation. *Blood.*102:972-980.

Yamada, M., Ikeuchi, T., Aimoto, S., and Hatanaka, H. (1996). PC12h-R cell, a subclone of PC12 cells, shows EGF-induced neuronal differentiation and sustained signaling. *J. Neurosci. Res., 43:* 355–364.

Yankulov KY, Bentley DL (1997). Regulation of CDK7 substrate specificity by MAT1 and TFIIH. *The EMBO J.*16:1638–1646.

Yasmin R, Williams RM, Xu M, Noy N.(2005).Nuclear import of the retinoid X receptor, the vitamin D receptor, and their mutual heterodimer. *J. Biol. Chem.* 280:40152-40160.

Yen A., Roberson MS, Varvayanis S, and Lee AT. (1998). Retinoic acid induced mitogen-activated protein (MAP)/extracellular signal regulated kinase (ERK) kinase-dependent MAP kinase activation needed to elicit HL-60 cell differentiation and growth arrest. *Cancer Research* 58: 3163-3172.

Yoshida H, Imamura T, Fujiki A, Hirashima Y, Miyachi M, Inukai T, Hosoi H. (2012). Post-transcriptional modulation of C/EBPα prompts monocytic differentiation and apoptosis in acute myelomonocytic leukemia cells. *Leu. Res.* 36:735-741.

Zanello LP, Norman AW. (2004). Rapid modulation of osteoblast ion channel responses by 1alpha,25(OH)2-vitamin D3 requires the presence of a functional vitamin D nuclear receptor. *Proc. Natl. Acad. Sci. USA.* 101:1589-1594.

Zhang J, Liu WL, Tang DC, (2002). Identification and characterization of a novel member of olfactomedin-related protein family, hGC-1, expressed during myeloid lineage development. Gene. 283:83–93.

Zelent A, Guidez F, Melnick A, Waxman S, Licht JD. (2001). Translocations of the RARalpha gene in acute promyelocytic leukemia. *Oncogene.* 20:7186-7203.

Zhang Y, Zhang J, and Studzinski GP. (2006). AKT Pathway Is Activated by 1, 25-Dihydroxyvitamin D3 and ParticipatesIn Its Anti-Apoptotic Effect and Cell Cycle Control in Differentiating HL60 Cells. *Cell Cycle* 5: 447-451.

Zhao K, Li X, Zhao Q, Huang Y, Li D, Peng Z, Shen W, Zhao J,Zhou Q, Chen Z, Sims PJ, Wiedmer T, and Chen G (2004a). Protein kinase C_ mediates retinoic acid and phorbolmyristate. acetate–induced phospholipid scramblase 1 gene expression: its role in leukemic cell differentiation. *Blood.* 104:3731-3738.

Zhao S, Konopleva M, Cabreira-Hansen M, Xie Z, Hu W, Milella M, Estrov Z, Mills GB, Andreeff M. (2004b). Inhibition of phosphatidylinositol 3-kinase dephosphorylates BAD and promotes apoptosis in myeloid leukemias. *Leukemia.*18:267-275.

Zhu XG, Hanover JA, Hager GL, Cheng SY. (1998). Hormone-induced translocation of thyroid hormone receptors in living cells visualized using a receptor green fluorescent protein chimera. *J. Biol. Chem.* 273:27058-27063.

In: Myeloid Cells   ISBN: 978-1-62948-046-6
Editor: Spencer A. Douglas   © 2013 Nova Science Publishers, Inc.

Chapter 2

PROLINE RICH HOMEODOMAIN (PRH/HHEX) PROTEIN IN THE CONTROL OF HAEMATOPOIESIS AND MYELOID CELL PROLIFERATION AND ITS POTENTIAL AS A THERAPEUTIC TARGET IN MYELOID LEUKAEMIAS AND OTHER CANCERS

*R. M. Kershaw[1], K. Gaston[2] and P-S. Jayaraman[1]**
[1]School of Immunity and Infection, College of Medical and Dental Sciences, University of Birmingham, Edgbaston, Birmingham, UK
[2]School of Biochemistry, University Walk, University of Bristol, Bristol, UK

ABSTRACT

Myelopoiesis occurs as a consequence of the interplay between transcription factors and growth factor dependent signalling pathways impacting on cell proliferation, cell survival and cell differentiation. It is equally well established that mis-regulation of transcription factors and/or growth factor receptor pathways contributes to disruption of myelopoeisis and the genesis of chronic and acute myeloid leukaemias. The Proline Rich Homeodomain protein (PRH, also known as HHEX) is a

* Corresponding author: p.jayaraman@bham.ac.uk.

transcription factor that is essential for haematopoiesis and it is expressed in haematopoietic stem cells and all myeloid lineages. PRH also plays a role in early embryonic development and in the formation of many organ systems. Significantly, there is compelling evidence that PRH is mis-regulated in both chronic myeloid leukaemia and some subtypes of acute myeloid leukaemia as well as in T-cell leukaemias and some solid cancers. This chapter reviews the functions of PRH in embryonic development, normal haematopoiesis and in myeloid leukaemias and focuses on the structure, localisation, transcriptional activity and post-translational modifications of PRH. In myeloid cells, PRH has been shown to be a transcriptional repressor of genes that regulate cell proliferation/survival including multiple genes in the VEGF signalling pathway. Thus PRH is a potent inhibitor of myeloid cell proliferation. Protein kinase CK2 is a stress responsive kinase with pleiotropic activities that promotes cell proliferation and its activity is often elevated in leukaemia and solid tumours. Phosphorylation of PRH by CK2 inhibits DNA binding by PRH and alleviates growth inhibition by PRH. Therapeutic treatment for CML (imatinib and dasatinib) also influences PRH phosphorylation and increases transcriptional repression of growth regulatory genes by PRH. Here we review the evidence which suggests that PRH may be an important therapeutic target in CML and other leukaemias, as well as in other cancers.

INTRODUCTION

Growth factors and transcription factors direct an ordered pattern of gene expression during myelopoiesis which results in the commitment and maturation of myeloid cells. All haematopoiesis takes place in the bone marrow of adults and when this tightly regulated process fails, abnormal or immature blood cells are formed, contributing towards the development of leukaemia. There are four major forms of leukaemia: acute myeloid leukaemia (AML), acute lymphoblastic leukaemia (ALL), chronic myeloid leukaemia (CML) and chronic lymphocytic leukaemia (CLL). Leukaemias are often considered to be diseases of childhood but they can develop at any age, only ALL occurs most commonly in children, with other leukaemias most commonly diagnosed in people older than 65. Myeloid leukaemias, which include AML and CML, are clonal diseases of haematopoietic stem or progenitor cells. Both genetic and epigenetic alterations can contribute to AML and CML by modifying cellular processes including self-renewal, proliferation and differentiation. As discussed in more detail later, in CML the Abelson murine leukaemia (ABL) gene is very often fused with the breakpoint cluster region (BCR) gene generating the oncoprotein BCR-ABL [1]. Leukaemogenic alterations such as mutations or translocations that produce fusion proteins mainly affect genes involved in signalling, tumour suppression, RNA maturation, epigenetic regulation and transcriptional regulation [2]. In AML, one of the most commonly affected and mutated group of genes encode homeobox transcription factors [3].

Homeobox genes contain a 180 base pair DNA sequence which encodes a highly conserved 60 amino acid DNA-binding homeodomain [4,5]. Many homeodomain proteins influence myeloid development and maturation and some of these proteins are implicated in the development of myeloid leukaemias. Human homeodomain proteins are grouped into 11 homeobox gene classes loosely defined by protein motifs surrounding the homeodomain. The Antennapedia (ANTP) class is strongly associated with haematopoietic differentiation. The ANTP class is divided into homeobox (HOX) and NK-like (NKL) subclasses [6]. A number of HOX genes are mis-expressed in leukaemia via chromosomal rearrangement, deletion, amplification, mutation, over-expression, down-regulation and promoter hypermethylation [3,7]. Over-expression of any one of several HOX genes results in long latency leukaemias, defined as occurring after 9 months of age in mice [8-10]. Abnormalities in upstream regulators of homeodomain transcription factors, such as caudal type homeobox 2 (CDX2), can lead to HOX over-expression [11,12], as can chromosomal rearrangements involving the mixed lineage leukaemia (Mll) gene, the product of which positively regulates the transcription of HOX genes [13]. HOX genes can also be de-regulated by chromosomal rearrangements leading to the formation of novel fusion proteins containing the HOX homeodomain and these are commonly found in leukaemic cells. For example, a frequent homeodomain protein fusion partner is the Nucleoporin 98 (NUP98) gene. The NUP98 protein is responsible for RNA and protein transport between the nucleus and cytoplasm [14]. At least 9 HOX genes have been found as fusions with NUP98, including HOXA9 [15]. HOXA9 for example is a well characterised regulator of haematopoiesis and AML patients with a NUP98-HOXA9 fusion have a poor prognosis [16].

Myeloid cells and B cells express PRH, a homeodomain containing transcription factor also known as human haematopoietically expressed homeobox (HHEX) [17]. The full length human PRH protein has been purified [18,19] and the role of PRH in development has been reviewed in detail [20]. The PRH protein is a member of the ANTP class of homeodomain proteins. However, whilst PRH is considered part of the NKL subclass of ANTP homeodomain proteins which contain a tinman (TN) motif N-terminal to the homeodomain [21], it lacks a conserved arginine at position 5 in the homeodomain that all other family members possess [22] as well as an NKL-specific domain downstream of the homeodomain, therefore, PRH can also be considered to be in its own subclass.

PRH STRUCTURE, LOCALISATION AND POST-TRANSCRIPTIONAL MODIFICATION

PRH Protein Structure

Human PRH is a 270 amino acid (aa) protein with a predicted molecular mass of 30kDa consisting of three domains: an N-terminal proline-rich domain (aa 1-136), the DNA binding homeodomain (aa 137-196) and an acidic C-terminal region (aa 197-270). [17] (Figure 1). The PRH sequence is conserved across different vertebrate species; in particular the homeodomain differs by only a single amino acid in human and mouse [23]. The 60 amino acid homeodomain allows PRH to recognise specific DNA sequences and regulate target genes [24]. The homeodomain consists of an N-terminal arm and 3 alpha helices which form a helix-loop-helix-turn-helix secondary structure. The N-terminal arm and the 3rd helix or recognition helix interact with DNA. The N-terminal arm is thought to wrap around the DNA and contact specific base pairs in the minor groove whilst the recognition helix interacts with base pairs in the major groove [25].

The N-terminal proline rich domain is critical for PRH oligomerisation [26] and it can repress transcription independently of the PRH homeodomain when it is fused to a heterologous DNA-binding domain [27,28]. Several proteins interact with the PRH N-terminal domain (aa 1-136) including members of the Groucho/TLE co-repressor family (aa 32-38) [29], the promyelocytic leukaemic (PML) protein (aa 50-115) [30], translation factor eukaryotic initiation factor 4E (eIF-4E) (aa 18-24) [31], proteasome subunit C8 [32] and protein kinase CK2 β subunit [33]. The PRH C-terminal domain is involved in transcriptional activation [34] and has been shown to interact with other transcription factors or co-activators including serum response factor (SRF) [35].

The homeodomain binds DNA at the consensus PRH-binding site: 5'-C/TA/TATTAAA/G-3' [17] however the recognition site is variable and short; the isolated PRH homeodomain has been shown to bind sequences containing

5'-TAAT-3', 5'-CAAG-3' or 5'-ATTAA-3' [36]. This flexibility combined with the absence of any obligatory binding partner proteins led researchers to question how PRH is able to bind specifically to its target sites. Some homeodomain containing proteins form dimers or larger oligomers when binding DNA [37-39], however PRH is unusual as it is capable of forming what appear to be homo-oligomeric octamers and double octamers (hexadecamers) both *in vitro* and in cells [26]. These oligomers are stable and highly resistant to unfolding by temperature and chemical denaturants [40]. The isolated homeodomain was shown to bind to a DNA fragment carrying a single PRH site with a 1:1 stoichiometry but the full length protein was shown to bind with a 1:7.2 stoichiometry [41]. It is therefore thought that PRH binds DNA as an octamer or hexadecamer, targeting genes with multiple PRH binding sites, as observed at the Goosecoid promoter [41]. This multimeric binding mechanism allows high binding specificity in the absence of a well-defined recognition site.

Interestingly, PRH oligomers can compact DNA and it is thought that this is a consequence of the binding of large PRH hexadecameric complexes to linear arrays of PRH sites [42]. The association of hexadecameric PRH-DNA complexes appears to allow PRH to spread along DNA. This could be similar to how members of the octameric leucine-responsive regulatory protein Lrp/AsnC family of proteins from bacteria and archaea bind DNA. The Lrp/AsnC proteins are involved in both transcriptional regulation and the global control of genome architecture but are thought to have no equivalents in eukaryotes [43,44]. As PRH and nucleosomes show similarities, in that they can interact with long stretches of DNA as well as with other protein partners, it will be important to determine whether PRH binds to nucleosomes or indeed replaces them when it binds to PRH binding site arrays.

PRH regulates genes transcriptionally both directly, by binding to gene promoters such as the endothelial specific molecule-1 (esm-1) promoter [45] and indirectly, by modulating the activity of other transcription factors such as members of the activator protein-1 (AP-1) family [46]. The ability of PRH to repress transcription relies on TLE co-repressor proteins, the human homologs of Groucho [29,42]. TLE proteins do not have DNA binding ability but are recruited by interactions with other DNA-binding proteins to form oligomeric complexes and bring about long-range transcriptional repression through interaction with histone deacetylases (HDACs) and histones [47-52]. PRH has been shown to bring about nuclear retention of endogenous TLE proteins in K562 cells, suggesting that some gene regulation could be via alteration of available nuclear TLE [53]. As described in detail below, PRH is also able to regulate genes post-transcriptionally by modifying the transport of specific mRNAs required for proliferation [31,54,55].

PRH Localisation

In embryonic endodermal cells that give rise to the liver PRH is nuclear but in cells lateral to the liver-forming region PRH is cytoplasmic [56]. However in many adult cells including K562 leukaemic cells, normal thyroid and breast cells, PRH is present in both the nucleus and cytoplasm [29,30,57,58]. In K562 cells, which are derived from a patient with chronic myeloid leukaemia (CML) in blast crisis [59], PRH is visible in foci in the nucleus and often colocalises with PML or eIF-4E nuclear bodies [29,30]. A mechanism for nucleo-cytoplasmic trafficking of PRH has been proposed involving nuclear import by Karyopherin/Importin 7 [60]. Interestingly, changes in PRH localisation correlate with neoplastic transformation; PRH is more confined to the cytoplasm in malignant thyroid tumours and in breast cancer cells [57,58].

PRH Post-Transcriptional Modification

Protein kinase CK2 is a ubiquitously expressed serine/threonine protein kinase which regulates multiple proteins involved in transcription, signalling, cell proliferation and DNA repair [61]. CK2 is a tetrameric complex consisting of two catalytic CK2α subunits and two regulatory CK2β subunits. It is considered to be oncogenic as CK2 promotes growth and inhibits apoptosis through negative regulation of tumour suppressors including p53 [62], PML [63] and PTEN (phosphatase and tensin homolog) [64] and stabilisation of oncogenes including c-Myc [65]. Levels of CK2 are elevated in many malignant cells [66,67] and high CK2 expression is associated with reduced patient survival [68-72]. The activity of transcription factors is commonly controlled by phosphorylation which can modulate DNA-binding affinity, subcellular compartmentalisation, stability and/or the ability to form protein-protein interactions [73]. PRH interacts with the β subunit of protein kinase CK2 which allows phosphorylation of PRH by the CK2α subunit [33]. PRH is phosphorylated in the homeodomain at serine residues 163 and 177 (Figure 1) and this inhibits the ability of PRH to bind DNA both *in vitro* and in cells [33,74]. As discussed in detail later CK2 also alters PRH subcellular localisation, decreases its stability and thereby inhibits PRH gene repression and growth control activities [74].

PRH IN EARLY DEVELOPMENT AND ORGANOGENESIS

Embryonic development requires PRH expression for correct patterning of the body axis and formation of the forebrain and head structures. Three functions of PRH appear crucial for development: 1) tissue-specific control of gene expression involving direct and/or indirect regulation of target genes, 2) promotion of expression of diffusible proteins that inhibit the action of signalling pathways in adjacent tissues, and 3) regulation of proliferation of precursor endodermal cells to allow endodermal progeny cells to lie outside a zone of inhibition and differentiate (reviewed by [20]). Mouse *Prh/Hex*[-/-] embryos begin to die 11.5 days *post coitum* (dpc) and none are present 15.5 dpc [75] hence PRH is critical for early development. PRH is expressed in a tissue-specific manner at different developmental stages and tissues including liver, lung, thymus and pancreas show high levels of PRH expression during embryogenesis while the lung, thyroid and liver retain high levels of PRH expression in adults [27,76-80].

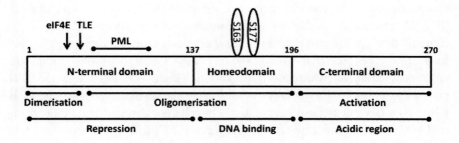

Figure 1. Domain structure of PRH. The three domains of PRH: N-terminal repression domain, DNA binding homeodomain and C-terminal acidic region domain. Interaction sites for eIF4E, TLE and PML as well as CK2-phosphorylation sites S163 and S177 are marked. The function of each region is also illustrated. Adapted from (20) and (157).

In the developing embryo the liver is an important site for haematopoiesis. Expression of PRH in the developing liver promotes organogenesis [77]. Activation of PRH transcription by HNF3 and GATA-4 results in the activation of several liver-specific genes by PRH [75,78,81,82]. PRH is required during multiple stages of hepatobiliary development and is essential for proper hepatoblast differentiation and bile duct development [83]. PRH is also expressed in adult liver but not in undifferentiated hepatocytes or undifferentiated cell lines [27]. Although the role of PRH in adult liver has not been well explored, a recent study has shown that in the adult liver the growth factor glypican 3 (GPC3) negatively regulates proliferation in part through PRH [84]. The binding of GPC3 to CD81 prevents CD81 binding to PRH thus freeing PRH to translocate to the nucleus and repress transcription [84].

Whilst the role of PRH in the developing thyroid [36,80], heart [85,86], skin [87] and pancreas [88,89] has been investigated, its role in these tissues in the adult has not been fully characterised. However, the expression of PRH in multiple organs throughout embryogenesis and into adulthood indicates that PRH plays a role in gene regulation at multiple stages of embryonic development as well as in the maintenance of mature organ function in the adult. Interestingly, a link between susceptibility to type 2 diabetes mellitus (T2DM) and polymorphisms in the *Prh* gene has been identified. In a genome-wide association screen single nucleotide polymorphisms (SNPs) in the IDE-KIF11-HHEX/PRH (insulin degrading enzyme, kinesin-interacting factor 11) locus were found to be significantly associated with T2DM [90]. A recent meta-analysis confirmed that two SNPs, rs1111875 and rs7923837, in close proximity and 3' to the *Prh* gene are significantly associated with diabetes [91]. This region 3' to the *Prh* gene contains a highly conserved non-coding element that can function as an enhancer element in reporter assays [92]. Interestingly, babies that inherit the rs1111875 SNP that lies within this enhancer element show a reduced foetal birth weight [93] and a high paediatric body mass index [94]. Population studies have shown that carriers of the PRH-IDE SNPs have decreased pancreatic beta-cell function and patients carrying rs1111875 show a low acute insulin response [95,96]. The reduced foetal birth weight may be due to decreased insulin secretion brought about by altered PRH levels in the developing foetus and this may correlate with increased susceptibility to T2DM in later life [93]. These studies together with models investigating pancreatic development in *Xenopus laevis* [89] and insulin resistance in *Drosophila melanogaster* [97] implicate alterations in PRH expression in human T2D.

PRH IN HAEMATOPOIETIC AND
VASCULAR DEVELOPMENT

Haematopoiesis in the developing embryo (Figure 2) is initiated in the blood islands of the yolk-sac. These extra-embryonic blood islands are formed from mesodermal progenitor haemangioblasts; multipotent precursor cells that consist of haematopoietic stem cells (HSCs) and angioblasts that can differentiate into both haematopoietic and endothelial cells [98,99]. As development progresses haematopoiesis occurs within the embryo, in the aorta gonad mesonephros (AGM) region and the foetal liver (Durand and Dzierzak 2005). Following birth and during all adult life all mature blood cells are produced in the bone marrow [98,99]. Tissues that contribute to both haematopoiesis and vasculogenesis express PRH during early embryonic development. Prior to gastrulation, murine PRH is expressed in extra-embryonic mesoderm in the blood islands of the yolk sac. After gastrulation, PRH is initially expressed in the embryo at sites where angioblasts are thought to arise and endocardial cells develop into the cardiac tubes of the heart, then later in development it is expressed in the foetal liver [56,77,79]. At sites of vessel formation, PRH expression is lost once cells have differentiated to form committed endothelial precursors and there is no PRH expression in any mature precursor cells of the endothelial lineage [79].

The role of PRH in haematopoiesis and vasculogenesis is complex and currently not fully understood. Studies using $Prh^{-/-}$ embryonic stem (ES) cells and embryonic bodies (EBs) report conflicting results. Embryonic stem cells are pluripotent culture cells that can contribute to all tissues after transplantation into a blastocyst embryo. When induced to differentiate *in vitro*, ES cells spontaneously differentiate into EBs, which contain precursor cells for various cell types including those for the haematopoietic lineages. Under appropriate culture conditions, some ES cells give rise to blast colonies in response to vascular endothelial growth factor (VEGF) and these cells generate both haematopoietic and endothelial cells and are considered equivalent to haemangioblasts [100]. Haemangioblasts give rise to both primitive and definitive haematopoiesis. Primitive haematopoiesis is defined as the production of progenitors which are unable to fully recapitulate haematopoiesis in an adult, whereas definitive haematopoiesis is defined as the production of progenitors able to fully recapitulate haematopoiesis in an adult [101].

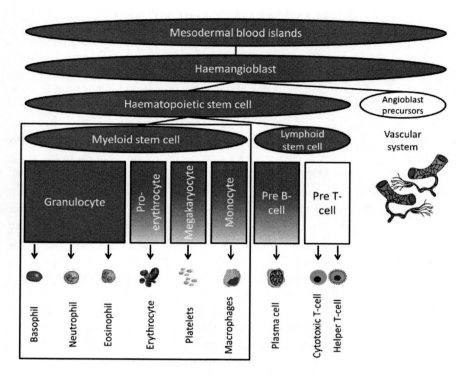

Figure 2. Haematopoiesis. Haematopoietic cells arise in the mesodermal blood islands. The haemangioblast gives rise to AGMs and angioblasts (vascular endothelial progenitors which form the vascular endothelial system). HSCs differentiate to both myeloid and lymphoid blood cells. PRH expression (shaded) is lost during terminal differentiation of all lineages (shaded region fades) except granulocytes. It is not expressed in T-cell lineages and angioblast precursors or their derivatives (white). Adapted from (20) and (157).

Using a (haemangio-) blast colony forming cell (BL-CFC) assay Guo et al. showed that lack of PRH affected definitive but not primitive embryonic haematopoiesis [101]. They showed that loss of PRH did not affect haemangioblast formation but that PRH is required for proper differentiation of the haemangioblast once it is formed. In contrast, Paz et al. used a murine ES cell/OP9 co-culture system to model embryonic haematopoiesis *in vitro* and showed that haematopoietic initiation coincides with increased PRH expression and that loss of PRH expression delays haemangioblast development, therefore PRH aids the initial formation of the haemangioblast [102]. Expression of PRH was not detected in the BL-CFCs at day 3 by Guo et al., however it was detected at the same time by Kubo et al., perhaps due to differences in sensitivity of the protocol or types of colonies assayed [101,103]. Additionally, whilst Guo et al. did not observe a difference in the number of BL-CFCs formed from day 3 $Prh^{+/+}$, $Prh^{+/-}$ and $Prh^{-/-}$ EBs, Kubo et al. showed that the frequency of BL-CFCs was increased in $Prh^{-/-}$ EBs and decreased when Prh was over-expressed [101,103]. Paz et al. propose that this increase observed by Kubo et al. is in keeping with their findings, suggesting that Prh functions as a negative regulator of haemangioblast development and may maintain proper numbers of progenitors [102]. Both studies showed that $Prh^{-/-}$ ES cells form fewer committed haematopoietic cells [101,103]. Although the data is conflicting and further clarification is required, overall the evidence suggests that PRH plays a role at different stages of development during EB differentiation, initially negatively regulating haemangioblast and endothelial proliferation and at later stages positively regulating endothelial cell maturation and haematopoietic differentiation.

Interestingly, the $Prh^{-/-}$ derived progenitors failed to transition into the more mature haematopoietic progenitors at the same rate and frequency as wild-type progenitors, accumulating in the G2 phase of the cell cycle [102]. An assessment of levels of cyclin D1, which is necessary for transition through the G1 phase revealed that cyclin D1 expression is down-regulated following loss of PRH expression [102]. This suggests that PRH may directly regulate the G1 phase of the cell cycle in the haematopoietic progenitor by up-regulating cyclin D1 expression. This is surprising given that studies in leukaemic cells show that PRH can inhibit expression of cyclin D1 by influencing its translation [31]. Cell lineage type differences must account for this difference in the effect of PRH on cyclin D1 expression. Regulation of the maturation of the definitive haematopoietic progenitor is mediated by several transcription factors and cell-cycle regulators [104]. One of these genes is the *Mll* gene [105], which regulates haematopoietic progenitor development partly through cell cycle regulation [106,107]. This gene encodes a DNA binding protein that also has enzymatic activity allowing it to methylate histone 3 lysine 4. Cytogenetic rearrangement of the *Mll* gene resulting in fusion of MLL with a variety of partner proteins is frequently associated with acute myeloid leukaemias [108]. The data presented by Paz et al. indicates that within the definitive haematopoietic progenitor a PRH-regulated signalling pathway may be driving expression of the *Mll* gene [102]. This is likely to be highly significant because mutations or changes in PRH will impact on this important epigenetic regulator that influences the expression of many of the Hox genes and regulates the Wnt signalling pathway.

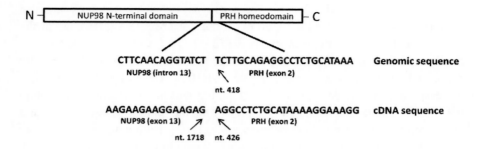

Figure 3. NUP98 fusion with PRH. NUP98 fuses in-frame with PRH. Nucleotide (nt.) 1718 of NUP98 joins nucleotide 426 of PRH. The genomic breakpoint takes place inside intron 13 in NUP98 and inside exon 2 in PRH resulting in joining of NUP98 intron 13 to nucleotide 418 of PRH exon 2. Adapted from (113).

As shown in Figure 2, PRH is expressed in early haematopoietic progenitors of all lineages except that of T-cells and its expression is generally down-regulated during terminal differentiation [23,109,110]. Terminally differentiated granulocytes are the exception where PRH expression is maintained [110]. Development of B-cells requires PRH [111] whereas development of T cells requires the down- regulation of PRH [112].

PRH AS A TUMOUR SUPPRESSOR IN AML AND CML

In myeloid lineages, PRH functions as an inhibitor of cell proliferation [31,103] and several studies have demonstrated mis-regulation of PRH and reduced PRH function in leukaemias of myeloid origin [31,113,114]. Notably Jankovic et al. characterised a NUP98/PRH fusion protein in an AML patient in relapse which transforms PRH from a strong transcriptional repressor to a transcriptional activator [113]. Chromosomal translocations involving the *NUP98* gene are linked with *de novo* and therapy-related AML and also found in ALL and advanced stages of CML [115]. In this NUP98/PRH fusion, the N-terminal domain of PRH is replaced by that of NUP98, with retention of the PRH homeodomain and therefore presumably preservation of DNA binding activity [113] (Figure 3). The region of NUP98 present in the fusion protein corresponds to the activation domain in NUP98 [116] and it seems likely that a transactivation domain has replaced the PRH repression and oligomerisation domain to allow activation of genes normally repressed by PRH. However, the fusion protein could also activate the transcription of multiple genes that are not normally repressed by PRH. Following transplantation of bone marrow cells into lethally irradiated syngenic Balb/C mice all NUP98/PRH mice developed an acute leukaemia involving the myeloid as well as the B-cell lineage with a latency period of 8 months [113]. This suggests that this translocation is not sufficient for leukaemia development however it demonstrates how mis-regulation of PRH can initiate a malignant phenotype in cells of the myeloid lineage.

Additionally, in several primary AML and blast crisis CML blood samples, Topisirovic et al. found PRH to be down-regulated and also mis-localised from the nucleus to the cytoplasm [114]. This coincided with up-regulation of protein levels of both cyclin D1 and eukaryotic translation initiation factor 4E (eIF4E) which, when over-expressed, leads to mis-regulated cellular proliferation and malignant transformation [117,118]. Increased cyclin D1 protein was shown to be a result of both increased transcription and increased eIF4E-dependent mRNA transport [114]. The transformation and growth promoting effects of eIF4E are inhibited by PRH, which interacts, as demonstrated by immunoprecipitation, with eIF4E in normal bone marrow cells but not CML [114]. Therefore in this context PRH acts to prevent acceleration of progression through the cell cycle caused by eIF4E, by inhibition of eIF4E-dependent mRNA transport of cyclin D1 [31]. Interestingly, cyclin D1 protein levels were up-regulated in all leukaemia specimens examined, indicating that mis-regulation of cyclin D1 is an important event in a variety of leukaemias [114].

The eIF4E protein also associates with PML tumour suppressor protein which negatively regulates eIF4E nuclear functions in response to cellular stress [119,120]. PML was first discovered through its involvement in acute promyelocytic leukaemia (APL). The PML gene fuses with the retinoic acid receptor α (RARα) gene to form PML-RAR and RARα-PML fusion proteins which drive the development of APL [121]. PML is involved in many cellular processes including cell cycle progression, DNA damage response and apoptosis. These activities require the localisation of PML to cellular organising centres termed PML nuclear bodies. These nuclear bodies are dynamic subnuclear structures and more than 100 cellular proteins are known to traffic in and out of them [122,123]. Intriguingly, both PRH and PML negatively regulate eIF4E activity and both can function as inhibitors of myeloid cell proliferation and inhibitors of angiogenesis [119,124-126]. Furthermore, PRH can interact directly with PML as well as with the PML-RAR fusion protein in APL cells. PML interacts with the N-terminal domain of PRH and PRH has been shown to be associated with PML nuclear bodies in

K562 and in APL patient-derived NB4 cells [30]. Moreover over-expression of PRH disrupts eIF4E and PML nuclear bodies by re-distributing them to the cytoplasm [31]. However, the role of the PML-PRH interaction in transcriptional repression or control of cell proliferation has not yet been investigated.

The involvement of PRH in tumour repression is not limited to haematopoietic cells as PRH has been implicated in cancers of the liver, breast and thyroid [57,58,127]. In both thyroid and breast carcinomas PRH expression was shown to be confined to the cytoplasm, presumably resulting in reduced capacity for gene regulation [57,58]. In hepatocellular carcinoma (HCC), the most common primary malignant tumour of the liver, a positive correlation was observed between PRH expression and the differentiation state of HCC clinical specimens; higher expression was observed in well-differentiated HCC specimens [127]. Furthermore, in support of a tumour suppressor role for PRH, PRH over-expression in a hepatoma cell line decreased the expression of multiple proto-oncogenes and increased the expression of some tumour suppressor genes including p53 and Rb. Moreover, the tumourigenicity of these cells in nude mice was attenuated following PRH over-expression [127].

PRH AS AN ONCOGENE IN ALL

Whilst PRH plays a tumour suppressor role in cells of myeloid origin, and appears to play a similar role in other tissues, in T cell lymphoid tissues PRH appears to be oncogenic. Human PRH is located on chromosome 10 near HOX11 and therefore may be dys-regulated in t(10;14) ALL when HOX11 is translocated to the T-cell receptor locus δ [128-131]. Mack et al. showed that T cell development in mice is dependent on down-regulation of PRH and that over-expression of PRH in transgenic mice results in abnormalities in T cell maturation, with increased immature myeloid cell proliferation [112]. Further to this, George et al. transplanted mice with bone marrow cells transduced with a retrovirus containing the PRH coding region and found that the recipient mice developed tumours originating in bone marrow from a precursor T-cell population [132]. Interestingly, these neoplasms are thought to develop due to inhibition of differentiation to mature T-cells. This demonstrates that PRH can act as an oncogene in lymphoid cells. Additionally, over-expression of PRH has been shown to be sufficient to initiate self-renewal of thymocytes *in vivo* and this is thought to be important for the accumulation of subsequent mutations resulting in leukaemic transformation [133]. The development of T-cell acute lymphoblastic leukaemia (T-ALL) can be induced by constitutive over-expression of the LMO2 oncogene, which forms a trimeric complex with the transcription factors ETS related gene (ERG) and friend leukaemia virus integration 1 (FLI1) to bind to the +1 enhancer of the PRH gene promoter [134]. McCormack et al. suggest that over-expression of LMO2 in thymocytes induces PRH expression, which can then initiate thymocyte self-renewal [133].

Interestingly Homminga et al. have noted that many members of the NKL homeodomain subclass are over-expressed in T cell leukaemias but not normally expressed in T-cell development [7]. They suggest that these T-ALL associated NKL homeobox genes might be providing a survival benefit by encoding proteins that up-regulate survival pathways present in a haematopoietic progenitor cell. They further suggest that the T-ALL NKL proteins might be 'mimicking' the normal function of PRH in maintaining early haematopoietic survival or self-renewal in cells where PRH is usually down regulated [7]. It is possible that this is also the situation when PRH is itself aberrantly up-regulated in T-cell leukaemias, such as in LMO2 dependent T-cell leukaemias. PRH could turn on self-renewal genes in committed T cell progenitors that would not normally be expressed at this stage of differentiation.

There is comparatively little evidence for the involvement of PRH in B cell leukaemias and lymphomas. However, using AKXD recombinant inbred mice, which develop leukaemias and lymphomas via retroviral activation of proto-oncogenes, PRH and another gene which flanks the viral insertion site, mEg5, a kinesin related protein necessary for spindle formation and stabilisation during mitosis, were shown to be up-regulated, suggesting that inappropriate PRH expression might contribute to B-cell leukaemias [135].

In light of the above it appears that PRH, like other NKL homeobox genes and many transcription factors can act as both oncogene and tumour suppressor gene depending on the cellular context.

PRH AND CK2 IN THE REGULATION OF K562 LEUKAEMIC CELL GROWTH

K562 cells are leukaemic blasts of myeloid lineage that retain the potential to spontaneously differentiate and can be manipulated to differentiate along several erythro-myeloid lineages [59,136]. They are a suitable model system for studying PRH in myeloid cells as they grow rapidly in culture, express high levels of PRH and are amenable to experimental PRH silencing and over-expression. Myeloid cell growth and survival is regulated by PRH, in part through direct transcriptional repression of genes in the VEGF signalling pathway (VSP) including VEGF and the VEGF receptors VEGFR-1, VEGFR-2 and neuropilin-1 [33,126]. The VSP is involved in multiple processes that result in tumour formation and progression to metastasis including angiogenesis, increased cell survival and haematopoieisis [137-140]. Elevated *Vegf* mRNA expression is seen in *Prh*[-/-] embryoid bodies, and *Prh*[-/-] knockout mice have increased VEGF protein [86,101]. The VEGF receptor genes are also down-regulated by PRH over-expression in human umbilical vein endothelial cells (HuVECs) [141]. By modifying PRH expression in CML blasts from the K562 cell line, Noy et al. showed that PRH binds directly to the promoter regions of the *Vegf*, *Vegfr-1* and *Vegfr-2* genes and represses their transcription [126]. Knockdown of PRH protein expression resulted in increased cell survival whereas PRH over-expression led to decreased survival due to increased apoptosis. These effects were shown to be mediated in part at least by changes in VEGF signalling [126].

Phosphorylation of PRH by CK2 inhibits PRH DNA binding activity and incubation of K562 cells with CK2 inhibitors decreases neuropilin-1, VEGFR-1 and VEGFA gene expression indicating less PRH phosphorylation and enhanced PRH repression activity [33,126]. The decrease in gene expression observed upon treatment of cells with CK2 inhibitors was not observed when PRH expression was silenced by shRNA knock-down. Conversely, the repression of VSP genes by PRH is alleviated by CK2 over-expression in K562 cells. Again, this alleviation was not observed when PRH expression was silenced, demonstrating that phosphorylation of PRH by CK2 regulates its transcriptional repression activity [74]. CK2 was also shown to antagonise the ability of over-expressed PRH to increase apoptosis, resulting in increased cell survival [74].

Phosphorylation of PRH by CK2 has a number of effects on the PRH protein including blocking the binding of PRH to the VEGFR-1 promoter, thereby preventing its transcriptional repression activity, altering the nuclear retention of PRH by rendering PRH less tightly held in the nucleus and altering PRH stability [33,74]. Using a phospho-PRH (pPRH) specific antibody Noy et al. showed that pPRH is less stable than non-phosphorylated PRH and is a substrate for cleavage by the proteasome. Cleavage of pPRH generates a C-terminally truncated PRH product of 27kDa [74]. Subsequent experiments revealed that this PRHΔC (p27) product can act as a transdominant negative regulator of PRH by sequestering TLE co-repressor proteins. CK2 phosphorylation can therefore inhibit the transcriptional activities of PRH in a variety of ways including the formation of a transdominant negative and this may be of clinical importance in malignant cells where CK2 activity is elevated.

PRH AS A THERAPEUTIC TARGET?

The ability of PRH to inhibit leukaemic cell growth and increase apoptosis indicates that it acts as a tumour suppressor in cells of myeloid origin and suggests that aberrant PRH activity is involved in tumour development. The inability of pPRH to exert this tumour suppressive activity suggests that increased pPRH relative to PRH is advantageous for these tumour cells. A mechanism for reducing pPRH, thus increasing non-phosphorylated transcriptionally active PRH, is therefore desirable and this could be achieved using a direct CK2 inhibitor such as CX-4945 (Silmitasertib) currently in phase I clinical trials for relapsed or refractory multiple myeloma, advanced solid tumours and Castleman's disease [142,143]. Alternatively an indirect inhibitor of CK2 such as Dasatinib, currently used clinically to treat CML, could be used to reduce pPRH [144].

The hallmark genetic abnormality of CML is a t(9;22) (q34;q11) translocation which generates the BCR-ABL fusion gene [145]. The BCR-ABL fusion protein is essential for initiation, maintenance and progression of CML but additional abnormalities are required for the transformation of CML from the initial chronic phase to the subsequent blast phase. The ABL protein is a non-receptor tyrosine kinase expressed in most tissues which regulates cytoskeleton structure [146] and the fusion of BCR to ABL increases the tyrosine kinase activity of ABL. The tyrosine kinase inhibitor imatinib mesylate (Gleevec) selectively induces apoptosis of BCR-ABL cells [147] and is used as a successful treatment for CML [148]. However residual BCR-ABL positive cells survive imatinib treatment and resistance emerges [149]. Dasatinib is a small molecule inhibitor of multiple tyrosine kinases including BCR-ABL and SRC family kinases and is used to treat newly diagnosed chronic-phase CML, or CML resistant to imatinib, as it is less susceptible to mechanisms of imatinib resistance (Quintás-Cardama and Cortes 2009).

In AML and CML CK2 activity is elevated [70,150]. CK2 has also been shown to be a target for ABL and BCR-ABL [151] and imatinib inhibition of

BCR-ABL in lymphoid cells inhibited CK2 activity [152]. Dasatinib treatment of K562 cells, a CML cell line which express BCR-ABL, was shown to inhibit cell survival [144]. Whilst CK2 is not a direct target of dasatinib or imatinib, both BCR-ABL inhibitors decreased PRH phosphorylation in K562 cells. Furthermore dasatinib treatment also increased VSP gene expression [144]. However, when PRH expression was down-regulated by shRNA treatment the reduction in pPRH and VSP gene expression by dasatinib was less pronounced and the effects of dasatinib on cell survival was greatly reduced. These findings indicate that increased CK2 activity as a consequence of BCR-ABL signalling is likely to result in more inactive PRH and increased VSP gene expression and this will contribute to increased survival. Hence either direct CK2 inhibition or more indirect inhibition of CK2 through targeting upstream Abl or Src kinases could be of therapeutic benefit (Figure 4).

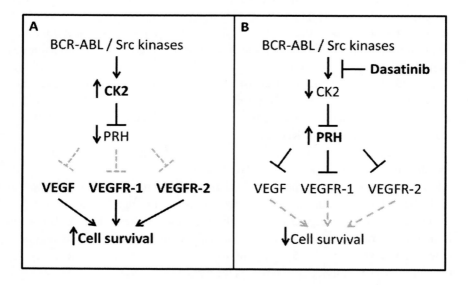

Figure 4. Mis-regulation of PRH activity in CML. A: Increased CK2 activity resulting from BCR-ABL/Src kinase activity leads to phosphorylation and inhibition of PRH, resulting in loss of PRH mediated VSP gene repression and increased cell survival. B: Dasatinib inhibits CK2 activity, leading to decreased PRH phosphorylation, re-established VSP gene repression and decreased cell survival. Adapted from (144).

Dasatinib inhibits multiple kinases however the data from Noy et al. suggests that PRH is required to mediate at least some of the effects of this inhibitor on cell survival [144]. Interestingly, multiple myeloma (MM) [153],

AML [70] and T-ALL [64] cells are all susceptible to apoptosis after treatment with CK2 inhibitors. In CML cells BCR-ABL signalling results in the constitutive expression of VEGF [154] and both VEGFR-1 and VEGFR-2 are highly expressed in the bone marrow of CML patients [155]. Mis-regulated VSP expression is associated with CML [156] hence there is a valid molecular basis for using CK2 inhibitors in conjunction with BCR-ABL inhibitors for treatment of primary CML.

CONCLUSION AND FUTURE PERSPECTIVES

The precise regulation of myeloid cellular proliferation and differentiation executed by homeodomain proteins is vital for haematopoiesis and their malfunction can result in leukaemias. In haematopoietic lineages where PRH is normally expressed the homeodomain protein PRH appears to function as a tumour suppressor, by inhibiting cell survival and growth. Although it remains to be determined whether this tumour suppressive ability is similar in other cell types, evidence for increased cytoplasmic localisation in breast and thyroid cells suggests that PRH is of importance in several different tumours. This renders it a potential therapeutic target in multiple tumour types. Decreased pPRH or increased hypo-phosphorylated PRH has been shown to inhibit myeloid tumour cell survival in culture. Using therapeutic agents which can alter pPRH and/or PRH levels, such as dasatinib, this laboratory research may be translated into patient treatments.

It is also possible that pPRH levels could be used as a biomarker for disease severity or prognosis, or to aid the choice of treatment for CML, AML and other diseases. Although phosphorylation of PRH by CK2 is the best characterised PRH post-translational modification, PRH is likely to undergo many other post–translational modifications which may be of importance for regulating PRH activity, subcellular localisation and stability. The identification of additional mechanisms for regulation of PRH activity/and or PRH expression will be important for discovering alternative methods for targeted tumour therapy and for tumour stratification and prognosis.

REFERENCES

[1] Rowley, J. D. (1973). *Nature*, *243*, 290-293.
[2] Murati, A., Brecqueville, M., Devillier, R., Mozziconacci, M. J., Gelsi-Boyer, V. & Birnbaum, D. (2012). *BMC Cancer*, *12*, 304
[3] Alharbi, R. A., Pettengell, R., Pandha, H. S. & Morgan, R. (2013). *Leukemia*, *27*, 1000-1008
[4] McGinnis, W., Levine, M. S., Hafen, E., Kuroiwa, A. & Gehring, W. J. (1984). *Nature*, *308*, 428-433
[5] Scott, M. P. & Weiner, A. J. (1984). *Proc Natl Acad Sci U S A*, *81*, 4115-4119.
[6] Holland, P. W., Booth, H. A. & Bruford, E. A. (2007). *BMC Biol*, *5*, 47
[7] Homminga, I., Pieters, R. & Meijerink, J. P. (2012). *Leukemia*, *26*, 572-581
[8] Kroon, E., Krosl, J., Thorsteinsdottir, U., Baban, S., Buchberg, A. M. & Sauvageau, G. (1998). *Embo J*, *17*, 3714-3725
[9] Sauvageau, G., Thorsteinsdottir, U., Hough, M. R., Hugo, P., Lawrence, H. J., Largman, C. & Humphries, R. K. (1997). *Immunity*, *6*, 13-22
[10] Thorsteinsdottir, U., Sauvageau, G., Hough, M. R., Dragowska, W., Lansdorp, P. M., Lawrence, H. J., Largman, C. & Humphries, R. K. (1997). *Molecular and Cellular Biology*, *17*, 495-505
[11] Rawat, V. P., Thoene, S., Naidu, V. M., Arseni, N., Heilmeier, B., Metzeler, K., Petropoulos, K., Deshpande, A., Quintanilla-Martinez, L., Bohlander, S. K., Spiekermann, K., Hiddemann, W., Feuring-Buske, M. & Buske, C. (2008). *Blood*, *111*, 309-319
[12] Scholl, C., Bansal, D., Dohner, K., Eiwen, K., Huntly, B. J., Lee, B. H., Rucker, F. G., Schlenk, R. F., Bullinger, L., Dohner, H., Gilliland, D. G. & Frohling, S. (2007). *J Clin Invest*, *117*, 1037-1048
[13] Slany, R. K. (2009). *Haematologica*, *94*, 984-993
[14] Gough, S. M., Slape, C. I. & Aplan, P. D. (2011). *Blood*, *118*, 6247-6257
[15] Borrow, J., Shearman, A. M., Stanton, V. P., Jr., Becher, R., Collins, T., Williams, A. J., Dube, I., Katz, F., Kwong, Y. L., Morris, C., Ohyashiki, K., Toyama, K., Rowley, J. & Housman, D. E. (1996). *Nat Genet*, *12*, 159-167
[16] Chou, W. C., Chen, C. Y., Hou, H. A., Lin, L. I., Tang, J. L., Yao, M., Tsay, W., Ko, B. S., Wu, S. J., Huang, S. Y., Hsu, S. C., Chen, Y. C., Huang, Y. N., Tseng, M. H., Huang, C. F. & Tien, H. F. (2009). *Leukemia*, *23*, 1303-1310

[17] Crompton, M. R., Bartlett, T. J., MacGregor, A. D., Manfioletti, G., Buratti, E., Giancotti, V. & Goodwin, G. H. (1992). *Nucleic Acids Res, 20*, 5661-5667

[18] Butcher, A. J., Gaston, K. & Jayaraman, P. S. (2003). *J Chromatogr B Analyt Technol Biomed Life Sci, 786*, 3-6

[19] Soufi, A., Gaston, K. & Jayaraman, P. S. (2006). *Int J Biol Macromol, 39*, 45-50

[20] Soufi, A. & Jayaraman, P. S. (2008). *Biochem J, 412*, 399-413

[21] Banerjee-Basu, S., & Baxevanis, A. D. (2001). *Nucleic Acids Res 29*, 3258-3269

[22] Neidle, S. & Goodwin, G. H. (1994). *FEBS Lett, 345*, 93-98

[23] Bedford, F. K., Ashworth, A., Enver, T. & Wiedemann, L. M. (1993). *Nucleic Acids Res, 21*, 1245-1249

[24] Gehring, W. J., Qian, Y. Q., Billeter, M., Furukubo-Tokunaga, K., Schier, A. F., Resendez-Perez, D., Affolter, M., Otting, G. & Wuthrich, K. (1994). *Cell, 78*, 211-223

[25] Billeter, M. (1996). *Prog Biophys Mol Biol, 66*, 211-225

[26] Soufi, A., Smith, C., Clarke, A. R., Gaston, K. & Jayaraman, P. S. (2006). *J Mol Biol, 358*, 943-962

[27] Tanaka, T., Inazu, T., Yamada, K., Myint, Z., Keng, V. W., Inoue, Y., Taniguchi, N. & Noguchi, T. (1999). *Biochem J, 339* (Pt 1)., 111-117

[28] Guiral, M., Bess, K., Goodwin, G., & Jayaraman, P. S. (2001). *J Biol Chem, 276*, 2961-2970

[29] Swingler, T. E., Bess, K. L., Yao, J., Stifani, S. & Jayaraman, P. S. (2004). *J Biol Chem, 279*, 34938-34947

[30] Topcu, Z., Mack, D. L., Hromas, R. A. & Borden, K. L. (1999). *Oncogene, 18*, 7091-7100

[31] Topisirovic, I., Culjkovic, B., Cohen, N., Perez, J. M., Skrabanek, L. & Borden, K. L. (2003). *Embo J, 22*, 689-703

[32] Bess, K. L., Swingler, T. E., Rivett, A. J., Gaston, K. & Jayaraman, P. S. (2003). *Biochem J, 374*, 667-675

[33] Soufi, A., Noy, P., Buckle, M., Sawasdichai, A., Gaston, K. & Jayaraman, P. S. (2009). *Nucleic Acids Res, 37*, 3288-3300

[34] Kasamatsu, S., Sato, A., Yamamoto, T., Keng, V. W., Yoshida, H., Yamazaki, Y., Shimoda, M., Miyazaki, J. & Noguchi, T. (2004). *J Biochem, 135*, 217-223

[35] Oyama, Y., Kawai-Kowase, K., Sekiguchi, K., Sato, M., Sato, H., Yamazaki, M., Ohyama, Y., Aihara, Y., Iso, T., Okamaoto, E., Nagai, R.

& Kurabayashi, M. (2004). *Arterioscler Thromb Vasc Biol*, *24*, 1602-1607

[36] Pellizzari, L., D'Elia, A., Rustighi, A., Manfioletti, G., Tell, G. & Damante, G. (2000). *Nucleic Acids Res*, *28*, 2503-2511

[37] Wilson, D. S., Guenther, B., Desplan, C. & Kuriyan, J. (1995). *Cell*, *82*, 709-719

[38] Palena, C. M., Chan, R. L. & Gonzalez, D. H. (1997). *Biochim Biophys Acta*, *1352*, 203-212

[39] Smith, D. L., Desai, A. B. & Johnson, A. D. (1995). *Nucleic Acids Res*, *23*, 1239-1243

[40] Shukla, A., Burton, N. M., Jayaraman, P. S. & Gaston, K. (2012). *PLoS One*, *7*, e35984

[41] Williams, H., Jayaraman, P. S. & Gaston, K. (2008). *J Mol Biol*, *383*, 10-23

[42] Soufi, A., Sawasdichai, A., Shukla, A., Noy, P., Dafforn, T., Smith, C., Jayaraman, P. S. & Gaston, K. (2010). *Nucleic Acids Res*, *38*, 7513-7525

[43] de los Rios, S. & Perona, J. J. (2007). *J Mol Biol*, *366*, 1589-1602

[44] Thaw, P., Sedelnikova, S. E., Muranova, T., Wiese, S., Ayora, S., Alonso, J. C., Brinkman, A. B., Akerboom, J., van der Oost, J. & Rafferty, J. B. (2006). *Nucleic Acids Res*, *34*, 1439-1449

[45] Cong, R., Jiang, X., Wilson, C. M., Hunter, M. P., Vasavada, H. & Bogue, C. W. (2006). *Biochem Biophys Res Commun*, 346, 535-545

[46] Schaefer, L. K., Wang, S. & Schaefer, T. S. (2001). *J Biol Chem*, *276*, 43074-43082

[47] Boddy, M. N., Freemont, P. S. & Borden, K. L. (1994). *Trends Biochem Sci 19*, 198-199

[48] Chen, G., Fernandez, J., Mische, S. & Courey, A. J. (1999). *Genes Dev*, *13*, 2218-2230

[49] Chen, G., Nguyen, P. H. & Courey, A. J. (1998). *Molecular and Cellular Biology 18*, 7259-7268

[50] Choi, C. Y., Kim, Y. H., Kwon, H. J. & Kim, Y. (1999). *J Biol Chem*, *274*, 33194-33197

[51] Palaparti, A., Baratz, A. & Stifani, S. (1997). *J Biol Chem*, *272*, 26604-26610

[52] Song, H., Hasson, P., Paroush, Z. & Courey, A. J. (2004). *Molecular and Cellular Biology*, *24*, 4341-4350

[53] Desjobert, C., Noy, P., Swingler, T., Williams, H., Gaston, K. & Jayaraman, P. S. (2009). *Biochem J*, *417*, 121-132

[54] Topisirovic, I. & Borden, K. L. (2005). *Histol Histopathol*, *20*, 1275-1284

[55] Sonenberg, N. & Gingras, A. C. (1998). *Curr Opin Cell Biol*, *10*, 268-275

[56] Ghosh, B., Ganea, G. R., Denson, L. A., Iannucci, R., Jacobs, H. C. & Bogue, C. W. (2000). *Pediatr Res*, *48*, 634-638

[57] D'Elia, A. V., Tell, G., Russo, D., Arturi, F., Puglisi, F., Manfioletti, G., Gattei, V., Mack, D. L., Cataldi, P., Filetti, S., Di Loreto, C. & Damante, G. (2002). *J Clin Endocrinol Metab*, *87*, 1376-1383

[58] Puppin, C., Puglisi, F., Pellizzari, L., Manfioletti, G., Pestrin, M., Pandolfi, M., Piga, A., Di Loreto, C. & Damante, G. (2006). *BMC Cancer*, *6*, 192

[59] Lozzio, C. B. & Lozzio, B. B. (1975). *Blood*, *45*, 321-334

[60] Ploski, J. E., Topisirovic, I., Park, K. W., Borden, K. L. & Radu, A. (2009). *Mol Cell Biochem*, *332*, 173-181

[61] Ahmed, K., Gerber, D. A. & Cochet, C. (2002). *Trends Cell Biol*, *12*, 226-230

[62] Allende-Vega, N., Dias, S., Milne, D. & Meek, D. (2005). *Mol Cell Biochem*, *274*, 85-90

[63] Scaglioni, P. P., Yung, T. M., Cai, L. F., Erdjument-Bromage, H., Kaufman, A. J., Singh, B., Teruya-Feldstein, J., Tempst, P. & Pandolfi, P. P. (2006). *Cell*, *126*, 269-283

[64] Silva, A., Yunes, J. A., Cardoso, B. A., Martins, L. R., Jotta, P. Y., Abecasis, M., Nowill, A. E., Leslie, N. R., Cardoso, A. A. & Barata, J. T. (2008). *J Clin Invest*, *118*, 3762-3774

[65] Channavajhala, P., & Seldin, D. C. (2002). *Oncogene*, 21, 5280-5288

[66] Tawfic, S., Yu, S., Wang, H., Faust, R., Davis, A. & Ahmed, K. (2001). *Histol Histopathol*, *16*, 573-582

[67] Trembley, J. H., Wang, G., Unger, G., Slaton, J. & Ahmed, K. (2009). *Cell Mol Life Sci*, *66*, 1858-1867

[68] Eddy, S. F., Guo, S., Demicco, E. G., Romieu-Mourez, R., Landesman-Bollag, E., Seldin, D. C. & Sonenshein, G. E. (2005). *Cancer Res*, *65*, 11375-11383

[69] Gapany, M., Faust, R. A., Tawfic, S., Davis, A., Adams, G. L. & Ahmed, K. (1995). *Mol Med 1*, 659-666

[70] Kim, J. S., Eom, J. I., Cheong, J. W., Choi, A. J., Lee, J. K., Yang, W. I. & Min, Y. H. (2007). *Clin Cancer Res*, *13*, 1019-1028

[71] Lin, K. Y., Fang, C. L., Chen, Y., Li, C. F., Chen, S. H., Kuo, C. Y., Tai, C. & Uen, Y. H. (2010). *Ann Surg Oncol*, *17*, 1695-1702

[72] P, O. c., Rusch, V., Talbot, S. G., Sarkaria, I., Viale, A., Socci, N., Ngai, I., Rao, P. & Singh, B. (2004). *Clin Cancer Res*, *10*, 5792-5803

[73] Whitmarsh, A. J. & Davis, R. J. (2000). *Cell Mol Life Sci*, *57*, 1172-1183

[74] Noy, P., Sawasdichai, A., Jayaraman, P.-S. & Gaston, K. (2012). *Nucleic Acids Res*, *40*, 9008-9020

[75] Martinez Barbera, J. P., Clements, M., Thomas, P., Rodriguez, T., Meloy, D., Kioussis, D. & Beddington, R. S. (2000). *Development*, *127*, 2433-2445

[76] Bogue, C. W., Ganea, G. R., Sturm, E., Ianucci, R. & Jacobs, H. C. (2000). *Dev Dyn*, *219*, 84-89

[77] Keng, V. W., Fujimori, K. E., Myint, Z., Tamamaki, N., Nojyo, Y. & Noguchi, T. (1998). *FEBS Lett*, *426*, 183-186

[78] Newman, C. S., Chia, F. & Krieg, P. A. (1997). *Mech Dev*, *66*, 83-93

[79] Thomas, P. Q., Brown, A. & Beddington, R. S. (1998). *Development*, *125*, 85-94

[80] Yatskievych, T. A. Pascoe, S., & Antin, P. B. (1999). *Mech Dev*, *80*, 107-109

[81] Denson, L. A., McClure, M. H., Bogue, C. W., Karpen, S. J. & Jacobs, H. C. (2000). *Gene*, *246*, 311-320

[82] Tanaka, H., Yamamoto, T., Ban, T., Satoh, S., Tanaka, T., Shimoda, M., Miyazaki, J. & Noguchi, T. (2005). *Arch Biochem Biophys*, *442*, 117-124

[83] Hunter, M. P., Wilson, C. M., Jiang, X., Cong, R., Vasavada, H., Kaestner, K. H. & Bogue, C. W. (2007). *Dev Biol*, *308*, 355-367

[84] Bhave, V. S., Mars, W., Donthamsetty, S., Zhang, X., Tan, L., Luo, J., Bowen, W. C. & Michalopoulos, G. K. (2013). *Am J Pathol.*

[85] Foley, A. C. & Mercola, M. (2005). *Genes Dev*, *19*, 387-396

[86] Hallaq, H., Pinter, E., Enciso, J., McGrath, J., Zeiss, C., Brueckner, M., Madri, J., Jacobs, H. C., Wilson, C. M., Vasavada, H., Jiang, X. & Bogue, C. W. (2004). *Development*, *131*, 5197-5209

[87] Obinata, A., Akimoto, Y., Omoto, Y. & Hirano, H. (2002). *Dev Growth Differ*, *44*, 281-292

[88] Bort, R., Martinez-Barbera, J. P., Beddington, R. S. & Zaret, K. S. (2004). *Development*, *131*, 797-806

[89] Zhao, H., Han, D., Dawid, I. B., Pieler, T. & Chen, Y. (2012). *Proc Natl Acad Sci U S A*, *109*, 8594-8599

[90] Sladek, R., Rocheleau, G., Rung, J., Dina, C., Shen, L., Serre, D., Boutin, P., Vincent, D., Belisle, A., Hadjadj, S., Balkau, B., Heude, B.,

Charpentier, G., Hudson, T. J., Montpetit, A., Pshezhetsky, A. V., Prentki, M., Posner, B. I., Balding, D. J., Meyre, D., Polychronakos, C. & Froguel, P. (2007). *Nature ,445*, 881-885

[91] Cai, Y., Yi, J., Ma, Y. & Fu, D. (2011). *Mutagenesis, 26*, 309-314

[92] Ragvin, A., Moro, E., Fredman, D., Navratilova, P., Drivenes, O., Engstrom, P. G., Alonso, M. E., de la Calle Mustienes, E., Gomez Skarmeta, J. L., Tavares, M. J., Casares, F., Manzanares, M., van Heyningen, V., Molven, A., Njolstad, P. R., Argenton, F., Lenhard, B. & Becker, T. S. (2010). *Proc Natl Acad Sci U S A, 107*, 775-780

[93] Freathy, R. M., Bennett, A. J., Ring, S. M., Shields, B., Groves, C. J., Timpson, N. J., Weedon, M. N., Zeggini, E., Lindgren, C. M., Lango, H., Perry, J. R., Pouta, A., Ruokonen, A., Hypponen, E., Power, C., Elliott, P., Strachan, D. P., Jarvelin, M. R., Smith, G. D., McCarthy, M. I., Frayling, T. M. & Hattersley, A. T. (2009). *Diabetes, 58*, 1428-1433

[94] Zhao, J., Bradfield, J. P., Zhang, H., Annaiah, K., Wang, K., Kim, C. E., Glessner, J. T., Frackelton, E. C., Otieno, F. G., Doran, J., Thomas, K. A., Garris, M., Hou, C., Chiavacci, R. M., Li, M., Berkowitz, R. I., Hakonarson, H. & Grant, S. F. (2010). *Diabetes, 59*, 751-755

[95] Grarup, N., Rose, C. S., Andersson, E. A., Andersen, G., Nielsen, A. L., Albrechtsen, A., Clausen, J. O., Rasmussen, S. S., Jorgensen, T., Sandbaek, A., Lauritzen, T., Schmitz, O., Hansen, T. & Pedersen, O. (2007). *Diabetes, 56*, 3105-3111

[96] Pascoe, L., Tura, A., Patel, S. K., Ibrahim, I. M., Ferrannini, E., Zeggini, E., Weedon, M. N., Mari, A., Hattersley, A. T., McCarthy, M. I., Frayling, T. M. & Walker, M. (2007). *Diabetes, 56*, 3101-3104

[97] Pendse, J., Ramachandran, P. V., Na, J., Narisu, N., Fink, J. L., Cagan, R. L., Collins, F. S. & Baranski, T. J. (2013). *BMC Genomics 14*, 136

[98] Baron, M. H. (2003). *Exp Hematol, 31*, 1160-1169

[99] Park, C., Ma, Y. D. & Choi, K. (2005). *Exp Hematol, 33*, 965-970

[100] Durand, C., & Dzierzak, E. (2005). *Haematologica, 90*, 100-108

[101] Guo, Y., Chan, R., Ramsey, H., Li, W., Xie, X., Shelley, W. C., Martinez-Barbera, J. P., Bort, B., Zaret, K., Yoder, M. & Hromas, R. (2003). *Blood, 102*, 2428-2435

[102] Paz, H., Lynch, M. R., Bogue, C. W. & Gasson, J. C. (2010). *Blood, 116*, 1254-1262

[103] Kubo, A., Chen, V., Kennedy, M., Zahradka, E., Daley, G. Q. & Keller, G. (2005). *Blood, 105*, 4590-4597

[104] Teitell, M. A. & Mikkola, H. K. (2006). *Pediatr Res, 59*, 33R-39R

[105] Ernst, P., Mabon, M., Davidson, A. J., Zon, L. I. & Korsmeyer, S. J. (2004). *Curr Biol*, *14*, 2063-2069

[106] Jude, C. D., Climer, L., Xu, D., Artinger, E., Fisher, J. K. & Ernst, P. (2007). *Cell Stem Cell*, *1*, 324-337

[107] McMahon, K. A., Hiew, S. Y., Hadjur, S., Veiga-Fernandes, H., Menzel, U., Price, A. J., Kioussis, D., Williams, O. & Brady, H. J. (2007). *Cell, Stem Cell 1*, 338-345

[108] Krivtsov, A. V. & Armstrong, S. A. (2007). *Nat Rev Cancer 7*, 823-833

[109] Jayaraman, P. S., Frampton, J. & Goodwin, G. (2000). *Leuk Res*, *24*, 1023-1031

[110] Manfioletti, G., Gattei, V., Buratti, E., Rustighi, A., De Iuliis, A., Aldinucci, D., Goodwin, G. H. & Pinto, A. (1995). *Blood*, 85, 1237-1245

[111] Bogue, C. W., Zhang, P. X., McGrath, J., Jacobs, H. C. & Fuleihan, R. L. (2003). *Proc Natl Acad Sci U S A*, *100*, 556-561

[112] Mack, D. L., Leibowitz, D. S., Cooper, S., Ramsey, H., Broxmeyer, H. E. & Hromas, R. (2002). *Immunology*, *107*, 444-451

[113] Jankovic, D., Gorello, P., Liu, T., Ehret, S., La Starza, R., Desjobert, C., Baty, F., Brutsche, M., Jayaraman, P. S., Santoro, A., Mecucci, C. & Schwaller, J. (2008). *Blood*, *111*, 5672-5682

[114] Topisirovic, I., Guzman, M. L., McConnell, M. J., Licht, J. D., Culjkovic, B., Neering, S. J., Jordan, C. T. & Borden, K. L. (2003). *Molecular and cellular biology*, *23*, 8992-9002

[115] Lam, D. H. & Aplan, P. D. (2001). *Leukemia*, *15*, 1689-1695

[116] Kalverda, B., Pickersgill, H., Shloma, V. V. & Fornerod, M. (2010). *Cell*, *140*, 360-371

[117] Lazaris-Karatzas, A., Montine, K. S. & Sonenberg, N. (1990). *Nature*, *345*, 544-547

[118] Lazaris-Karatzas, A. & Sonenberg, N. (1992). *Molecular and cellular biology*, *12*, 1234-1238

[119] Cohen, N., Sharma, M., Kentsis, A., Perez, J. M., Strudwick, S. & Borden, K. L. (2001). *Embo J*, *20*, 4547-4559

[120] Topisirovic, I., Capili, A. D. & Borden, K. L. (2002). *Molecular and cellular biology*, *22*, 6183-6198

[121] de The, H., Lavau, C., Marchio, A., Chomienne, C., Degos, L. & Dejean, A. (1991). *Cell*, *66*, 675-684

[122] Lallemand-Breitenbach, V. & de The, H. (2010). *Cold Spring Harb Perspect Biol*, *2*, a000661

[123] Reineke, E. L. & Kao, H. Y. (2009). *Int J Biol Sci*, *5*, 366-376

[124] Bernardi, R., Guernah, I., Jin, D., Grisendi, S., Alimonti, A., Teruya-Feldstein, J., Cordon-Cardo, C., Simon, M. C., Rafii, S. & Pandolfi, P. P. (2006). *Nature, 442*, 779-785

[125] Kentsis, A., Dwyer, E. C., Perez, J. M., Sharma, M., Chen, A., Pan, Z. Q. & Borden, K. L. (2001). *J Mol Biol, 312*, 609-623

[126] Noy, P., Williams, H., Sawasdichai, A., Gaston, K. & Jayaraman, P.-S. (2010). *Molecular and Cellular Biology, 30*, 2120-2134

[127] Su, J., You, P., Zhao, J. P., Zhang, S. L., Song, S. H., Fu, Z. R., Ye, L. W., Zi, X. Y., Xie, D. F., Zhu, M. H. & Hu, Y. P. (2012). *Med Oncol, 29*, 1059-1067

[128] Li, J., Shen, H., Himmel, K. L., Dupuy, A. J., Largaespada, D. A., Nakamura, T., Shaughnessy, J. D., Jr., Jenkins, N. A. & Copeland, N. G. (1999). *Nat Genet, 23*, 348-353

[129] Hatano, M., Roberts, C. W., Minden, M., Crist, W. M. & Korsmeyer, S. J. (1991). *Science, 253*, 79-82

[130] Hromas, R., Radich, J. & Collins, S. (1993). *Biochem Biophys Res Commun, 195*, 976-983

[131] Kennedy, M. A., Gonzalez-Sarmiento, R., Kees, U. R., Lampert, F., Dear, N., Boehm, T. & Rabbitts, T. H. (1991). *Proc Natl Acad Sci U S A, 88*, 8900-8904

[132] George, A., Morse, H. C., 3rd, & Justice, M. J. (2003). *Oncogene, 22*, 6764-6773

[133] McCormack, M. P., Young, L. F., Vasudevan, S., de Graaf, C. A., Codrington, R., Rabbitts, T. H., Jane, S. M. & Curtis, D. J. (2010). *Science, 327*, 879-883

[134] Oram, S. H., Thoms, J. A., Pridans, C., Janes, M. E., Kinston, S. J., Anand, S., Landry, J. R., Lock, R. B., Jayaraman, P. S., Huntly, B. J., Pimanda, J. E. & Gottgens, B. (2010). *Oncogene, 29*, 5796-5808

[135] Hansen, G. M. & Justice, M. J. (1999). *Oncogene, 18*, 6531-6539

[136] Drexler, H. G. (1994). *Leuk Res, 18*, 919-927

[137] Dvorak, H. F., Brown, L. F., Detmar, M. & Dvorak, A. M. (1995). *Am J Pathol, 146*, 1029-1039

[138] Gerber, H. P., Malik, A. K., Solar, G. P., Sherman, D., Liang, X. H., Meng, G., Hong, K., Marsters, J. C. & Ferrara, N. (2002). *Nature, 417*, 954-958

[139] Hattori, K., Heissig, B., Wu, Y., Dias, S., Tejada, R., Ferris, B., Hicklin, D. J., Zhu, Z., Bohlen, P., Witte, L., Hendrikx, J., Hackett, N. R., Crystal, R. G., Moore, M. A., Werb, Z., Lyden, D. & Rafii, S. (2002). *Nat Med, 8*, 841-849

[140] Wu, Y., Hooper, A. T., Zhong, Z., Witte, L., Bohlen, P., Rafii, S. & Hicklin, D. J. (2006). *Int J Cancer*, *119*, 1519-1529

[141] Nakagawa, T., Abe, M., Yamazaki, T., Miyashita, H., Niwa, H., Kokubun, S. & Sato, Y. (2003). *Arterioscler Thromb Vasc Biol*, *23*, 231-237

[142] National Cancer Institute. (2012). Clinical Trials, Dose-escalation Study of Oral CX-4945

[143] National Cancer Institute. (2012). Clinical Trials, Study of CX-4945 in Patients With Relapsed or Refractory Multiple Myeloma

[144] Noy, P., Gaston, K. & Jayaraman, P. S. (2012). *Leuk Res 36*, 1434-1437

[145] Ren, R. (2005). *Nat Rev Cancer*, *5*, 172-183

[146] Woodring, P. J., Hunter, T. & Wang, J. Y. (2003). *J Cell Sci*, *116*, 2613-2626

[147] Druker, B. J., Tamura, S., Buchdunger, E., Ohno, S., Segal, G. M., Fanning, S., Zimmermann, J. & Lydon, N. B. (1996). *Nat Med*, *2*, 561-566

[148] Druker, B. J. (2002). *Oncogene*, *21*, 8541-8546

[149] Lowenberg, B. (2003). *N Engl J Med*, *349*, 1399-1401

[150] Phan-Dinh-Tuy, F., Henry, J., Boucheix, C., Perrot, J. Y., Rosenfeld, C. & Kahn, A. (1985). *Am J Hematol*, *19*, 209-218

[151] Heriche, J. K. & Chambaz, E. M. (1998). *Oncogene*, *17*, 13-18

[152] Mishra, S., Reichert, A., Cunnick, J., Senadheera, D., Hemmeryckx, B., Heisterkamp, N. & Groffen, J. (2003). *Oncogene*, *22*, 8255-8262

[153] Piazza, F. A., Ruzzene, M., Gurrieri, C., Montini, B., Bonanni, L., Chioetto, G., Di Maira, G., Barbon, F., Cabrelle, A., Zambello, R., Adami, F., Trentin, L., Pinna, L. A. & Semenzato, G. (2006). *Blood*, *108*, 1698-1707

[154] Mayerhofer, M., Valent, P., Sperr, W. R., Griffin, J. D. & Sillaber, C. (2002). *Blood*, *100*, 3767-3775

[155] Verstovsek, S., Lunin, S., Kantarjian, H., Manshouri, T., Faderl, S., Cortes, J., Giles, F. & Albitar, M. (2003). *Leuk Res*, *27*, 661-669

[156] Janowska-Wieczorek, A., Majka, M., Marquez-Curtis, L., Wertheim, J. A., Turner, A. R. & Ratajczak, M. Z. (2002). *Leukemia 16*, 1160-1166

[157] Noy, P. (2011). Coordinate regulation of VEGF signalling genes by the proline rich homeodomain protein. in *Immunity and Infection*, University of Birmingham, Birmingham

In: Myeloid Cells
Editor: Spencer A. Douglas

ISBN: 978-1-62948-046-6
© 2013 Nova Science Publishers, Inc.

Chapter 3

APOPTOSIS, CELL CYCLE AND EPIGENETIC PROCESSES DEREGULATION IN MYELOPROLIFERATIVE NEOPLASMS

Daniela Dover de Araujo[*1], *Natalia de Souza Nunes*[2], *Raquel Tognon-Ribeiro*[3], *Sandra Mara Burin*[4] *and Fabiola Attié de Castro*[†5]

[1]Clinical Analyses, Toxicology and Food Sciences Department, School of Pharmaceutical Science of Ribeirão Preto-University of São Paulo, Ribeirão Preto, Brazil
University of Educational Foundation of Guaxupé- UNIFEG, Minas Gerais, Brazil
[2,3,4,5]Clinical Analyses, Toxicology and Food Sciences Department, School of Pharmaceutical Science of Ribeirão Preto-University of São Paulo, Ribeirão Preto, Brazil

ABSTRACT

Myeloproliferative neoplasms (MPN) are hematological diseases characterized by myeloproliferation/myeloaccumulation of mature cells without a specific stimulus. According to the World Health Organization

[*] E-mail: danydover@hotmail.com
[†] E-mail: castrofa@fcfrp.usp.br

(WHO, 2008), MPN are composed by Philadelphia (Ph) chromosome negative MPN, including Primary Myelofibrosis (PMF), Essential Thrombocythemia (ET), Polycythemia Vera (PV), uncommon MPN as well as chronic neutrophilic leukemia, mastocytosis and by Chronic Myeloid Leukemia (CML), a Ph positive chromosome MPN. In this chapter we focus on Primary Myelofibrosis (PMF), Essential Thrombocythemia (ET), Polycythemia Vera (PV) and Chronic Myeloid Leukemia (CML) pathogenesis, specifically apoptosis, cell cycle and epigenetic processes deregulation.

The JAK2 V617F mutation, which leads to constitutive JAK2 tyrosine kinase activation, is found in 95% of PV patients and in at least 50% of ET and PMF patients. JAK2 constitutive enzymatic activation is linked to prolonged cell survival and myeloproliferation. Other mutations, recently described, such as in JAK2 exon 12 and in the TET2, CBL, MPL and AXSL genes may also contribute to ET, PV and PMF pathogenesis and progression to acute myeloid leukemia. Some acquired genetic lesions in molecules involved in cytokine signaling and epigenetic regulation are also implicated in Ph negative MPN.

In CML patients, the expression of BCR-ABL1 determines the leukemogenesis process by increasing cell proliferation, promoting apoptosis impairment and deregulation of cell adhesion to bone marrow stroma and altering the cell epigenetic pattern.

Elucidation of cellular and molecular mechanisms involved in MPN pathogenesis is relevant for new therapeutic targets discovery and for description of potentials diagnostic and prognostic markers for diseases.

INTRODUCTION: DEFINITION, CLASSIFICATION AND CLINIC-LABORATORIAL FINDINGS OF MYELOPROLIFERATIVE NEOPLASMS

According to the 2008 World Health Organization (WHO) classification of Myeloid Neoplasms and Acute Leukemia, the term "myeloid" refers to all cells belonging to the granulocytic (neutrophil, eosinophil, basophil), monocytic/macrophage, erythroid, megakaryocytic and mast cell lineages. Among eight categories listed by WHO, Myeloproliferative Neoplasms is one of them, and comprises chromosome (Cr) Philadelphia (Ph) positive (Chronic Myeloid Leukemia, CML) and Ph (-) MPN or chronic MPN (Polycythemia Vera-PV, Essential Thrombocythemia-ET, and Primary Myelofibrosis-PMF) besides Chronic neutrophilic leukemia, Chronic eosinophilic leukemia not otherwise specified; Mastocytosis and Myeloproliferative neoplasms unclassifiable. The nomenclature for the myeloproliferative entities was changed from "chronic myeloproliferative disease" to "myeloproliferative neoplasms" in order to accurately reflect their neoplastic nature. [1]

CML results from clonal expansion of hematopoietic precursors cell. [2] The CML primarily affects adults between 50 and 60 years old, with annual incidence of 1 or 2 cases per 100,000 inhabitants per year, with a slight male predominance. [3] Its natural history is slowly progressive, where about 30% of cases are asymptomatic, and those presenting signs and symptoms, usually have a persistent anemia, associated with weakness, weight loss, fever and even abdominal pain. Associated with these symptoms, early diagnosis is accomplished through routine laboratory tests such as complete blood count, and presents a marked leukocytosis with a left shift.

The WHO established the criteria for the CML three phases: chronic, accelerated and blast phases. In the chronic phase most patients are asymptomatic or present nonspecific signs and symptoms such as fatigue, anorexia and sweating. Some authors consider that the chronic phase has two stages, the "initial chronic phase", which the BCR-ABL1 positive clones are expanding inside the bone marrow and differentiate into mature cells, and blasts are not detected in the peripheral blood. [4] At this stage, some lymphoid and myeloid cells circulating hold the BCR-ABL oncogene. [2, 5] The late chronic phase of the leukemic cells may present SH1 domain mutations of the ABL portion, as well as breaks in DNA derived from cytogenetic abnormalities acquired after or without treatment, but still with clinical manifestations of chronic phase [4].

The accelerated phase may occur months or years after disease diagnosis, depending on the patient's response to treatment in the chronic phase. Cytogenetic abnormalities mark this transition, in which the leukemic clones respond to treatment, especially due to mutations in the ATP binding site of the SH1 domain of BCR-ABL [5]

The blast phase is characterized by increased blasts (> 20%) in bone marrow and peripheral blood leukocytosis, anemia and thrombocytopenia. [6, 7] At this stage, failure occurs in the maturation of malignant myeloid precursors, which often have additional cytogenetic changes leading to the worst CML. [8, 9] Despite the knowledge about the effect of BCR-ABL1 in cell transformation, the cellular and molecular mechanisms responsible for the onset and progression of the disease to accelerate and blastic phases have not been fully elucidated. [10]

For MPN Ph (-) (PV, ET and PMF) the criteria used to initial diagnosis should be observed in peripheral blood (PB) and bone marrow (BM) specimens. Morphologic, cytochemical, and/or immunophenotypic features are used by WHO to guide the diagnosis since the first MPN classification, but in 2008 the information about JAK2 V617F mutation and similar activating mutation were included in the diagnostic algorithms for PV, ET, and PMF. Additional clinical, laboratory, and histologic parameters have been also included to allow diagnosis and subclassification regardless of whether JAK2 V617F or a similar mutation is present or not [1].

Individually PV and ET are characterized by an increased count of red blood cells and platelets respectively, while the PMF initial phase is characterized by increased granulocyte and megakaryocyte alteration responsible for increased platelet production. All the MPN patients are predisposed to thrombohemorrhagic events and vascular complications and can suffer from splenomegaly and other symptoms such as pruritus, night sweats, fatigue, and bone pain (severity vary according to the disease). [11, 12] MPN can develop a fibrotic phase, in PV and ET this phase is named 'post polycythemia vera myelofibrosis' (Post-PV MF) or 'post-essential thrombocythemia myelofibrosis (Post-ET MF), respectively. The progression of fibrotic phase is associated with anemia, splenomegaly, peripheral blasts and decreased survival. The most common complications of MPNs involve progressive cytopenias, weight loss, and blastic transformation. [13] Nowadays the available therapies assist an improving symptom but do not reverse the bone marrow fibrosis [12].

MYELOPROLIFERATIVE NEOPLASMS PHYSIOPATHOLOGY: MOLECULAR ASPECTS

Genetic and epigenetic alterations are a hallmark of MPN. Many mutations and cytogenetics alterations have been described and have being recognized as important for the physiopathology, and establishment of diagnostic and prognostic markers in myeloid neoplasms.

The first description of a cytogenetic abnormality associated with a malignant disease process was the Ph Cr in CML. [14, 15]. This cytogenetic marker, described in 1960 by Nowell and Hungerford is detected in 95% of CML patients. The Ph Ch originates from a reciprocal translocation of Ch 9 and 22 (q34, q11), culminating in the emergence of BCR-ABL1 neogene. [16] The *BCR-ABL1* encodes the BCR-ABL oncoprotein, which presents a constitutive tyrosine kinase (TK) activity. The oncoprotein BCR-ABL plays an uncontrolled TK activity and phosphorylates several substrates which are responsible to activate multiple signaling pathways, such as Ras, signal transducer and activator of transcription-5 (STAT-5), extracellular signal-regulated kinase (ERK)/mitogen-activated protein kinase (MAPK), Janus kinase 2 (JAK-2), phosphatidylinositol-3 kinase (PI-3K) and nuclear factor (NF)-κB. [2, 17] This deregulated signaling leads to the malignant cellular phenotype of CML, including increased cell proliferation, inhibition of apoptosis and reduction of cell adhesion to the bone marrow stroma and extracellular matrix. [18].

Other secondary chromosomal abnormalities may be found in the CML patient, such as trisomy 8 (+8; 34% of cases with additional changes), +Ph (30%), i(17q) (20%), +19 (13%) and they usually configure the clonal disease evolution. [19]

The Ph (-) MPN, the JAK2 V617F mutation (G → T, 1849 position, éxon 14), described in 2005, leads to constitutive activation of JAK2 tyrosine kinase. [20-24] The cells show hypersensitive to cytokines and hyperactivated STAT5, MAPK/ERK e PI3K/AKT pathways. [25] This mutation has been associated with more complicated clinical features, such as high hematocrit and hemoglobin levels, low erytropoetin (EPO) level and low white blood cell and neutrophil counts, as well as splenomegaly and thrombotic events. [26, 27]

Under physiological conditions JAK2 is in inactive conformation and noncovalently bound to cytokine class I receptors, important for myelopoiesis (EpoR, TpoR, G-CSFR), via its N-terminal FERM domain. Binding of ligands to their receptors (like erythropoietin, thrombopoietin) promotes a change in the JAK2 conformation and activates JAK2 signaling through tyrosine residue phosphorylation resulting in a pathway cascade activation including MAPK, STAT and PI3K proteins. [11] Mutations in both exons of JAK2 (éxon14 and éxon 12) promote constitutive activation of the protein even in the absence of cytokines.

However, the JAK V617F mutation is present in about 95% of PV patients but only 50% of ET and PMF patients. This finding indicates that it is likely that another genetic lesion may precede the JAK2 mutation and lead to the MPN diseases. [28] Given that, numerous studies have been carried out around the search for other changes involved in this process. The last 7 years changed the perspective of mutations described as involved in Ph (-) MPN, such as TET2, CBL, MPL, AXSL, EZH2, IDH, IKZF and LNK genes may contribute to these diseases pathogenesis and progression as well as acquired genetic lesions in molecules involved in cytokine signaling and epigenetic regulation. [29]

Despite the low incidence, these mutations play an important role in Ph (-) MPN pathophysiology. Mutations in the thrombopoetin receptor MPL were found in around 15% of JAK2 V617F negative- ET and PMF, this mutation occurs in exon 10, the region responsible for prevent the TpoR spontaneous activation. [29, 30] The adaptor protein LNK is a negative regulator of JAK2 signaling and it has an important role in hematopoiesis. Mutations in the LNK gene have been described in a frequency of about 10% after progressing to leukemia and therefore, this finding has been associated with disease progression. [31] Another important player in JAK regulation is the CBL, a ubiquitin ligase protein responsible for inducing tyrosine kinase proteossomal degradation (i.e., JAK2) and it has been described as mutated in 10% of PMF patients. [11, 32, 33] Therefore, it is evident that other mutations contribute to the alteration in JAK2 tyrosine kinase function and consequently for the development of Ph (-) MPN.

Apoptosis: Concepts and Alterations in MPN

Apoptosis or programmed cell death is an essential process for the maintenance of cellular homeostasis in living beings. The abnormal resistance to apoptosis can lead to the onset of cancer and autoimmune diseases, while the exacerbation leads to the onset of neurodegenerative diseases. [34] The hypothesis that alterations in apoptosis regulation could be involved in neoplastic processes in MPN was formed in the 1990s.

The apoptosis process in cells causes morphological characteristics, which begin with the loss of adhesion or extracellular matrix and with neighboring cells, causing cell shrinkage, chromatin condensation concomitantly present, DNA fragmentation, cell membrane and bubbles formation of apoptotic bodies. [35] These morphological changes are the result of a cascade of molecular and biochemical events specific and genetically regulated. Given a stimulus, apoptosis triggers several molecular events that culminate in the activation of proteases family members called caspases. [36] The caspase cascade activation can be triggered by either the intrinsic or extrinsic pathways.

The mitochondrial or intrinsic pathway begins with the action of different intracellular stress signals such as chemotherapeutic agents, ATP deficiency, cytoskeletal structures interruption, irradiation, viruses, bacteria and absence of cellular growth factors, which converge to mitochondria. The activation of this pathway depends on the release of cytochrome-c, from the mitochondria to the cytosol; this molecule interacts with APAF-1, forming the apoptosome and promotes the activation of caspase-9, which in turn activates caspase-3, -6 and -7 generating the apoptotic phenotype. [35]

Some molecules that act on mitochondria are central to the control of apoptosis by the intrinsic pathway, such as proteins of the BCL-2 family members, which inhibits apoptosis (BCL-X_L, MCL-1, A1, BCL-2 AND BCL-W) and promotes or sensitizes the cell to the cell death (BAD, BAX, BAK, BID, etc.). [37] (Figure 1)

The extrinsic or death receptor is first activated by the interaction of receptors located on the cell surface called "death receptors" to specific ligands.

These receptors are members of the receptor family of tumor necrosis factor (TNF), which contains a death domain essential for the transfer of apoptotic signals, also including FAS, TNFR1, DR3-6, among others. This pathway is initiated by the trimerization of the death receptors, which, for example, in the case of Fas-FasL interaction, allows the recruitment of two signaling proteins, FADD and PRO-CASPASE-8, forming the complex DISC (Death-Inducing Signaling Complex). The pro-caspase-8 is activated and released into the cytoplasm, where they act on stage performing the process to activate caspase-3, or promote the triggering of the intrinsic pathway by molecule cleavage BID. [37]

The recruitment and activation of caspase-8 are targets of anti-apoptotic proteins such as c-Flip. [38] (Figure 1)

The CML malignant hematopoietic precursor cells are more resistant to apoptosis than normal cells because of the expression TK BCR-ABL, which prolongs cell survival. [16, 39]

The molecular and cellular mechanisms involved in BCR-ABL + cells resistance to apoptosis is still controversial. Some studies report that is related to increased expression of BCL-X_L, and not of BCL-2. (36, 40) On the other hand, HANDA and colleagues (1997) reported that, in CML, the expression of BCL-2 protein increases with the disease progression, which supports the hypothesis that it is the expression of this protein at high levels prolonging survival and the leukemic cell becomes resistant to apoptosis.

Other reports using cell lines BCR-ABL+ showed that the oncoprotein interferes in the survival of the leukemic cell by deregulation of molecules involved in apoptosis control, such as reducing the expression of pro-apoptotic proteins BIM, FASL, P53 and BAX, increasing the expression of anti-apoptotic proteins such as SURVIVIN and MCL-1 [41-43].

In 1998, Silva and colleagues found high expression levels of anti-apoptotic protein BCL-X$_L$ in erythroid precursors of PV patients. This raises the expression associated with the capacity of the cells to differentiate in absence of EPO and improved resistance to apoptosis.

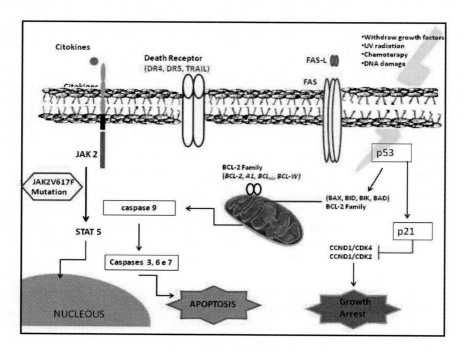

Figure 1. The Death receptor (Extrinsic) and Mitochondrial (Intrinsic) apoptosis pathways and the cell cycle components.

Only in 2004, Zhang et al. described BCL-X$_L$ alterations in ET, decreased gene expression levels found corroborate with the alterations observed in platelets production due to the important involvement of the BCL-X$_L$ gene in megakaryocytopoiesis. [44]

Alterations in the genes *FAS*, *FASL*, *FAIM*, *C-FLIP*, *TRAIL*, *DR4* and *DR5* receptors, components of the death receptor pathway were described in PV, ET and PMF by Tognon et al. (2011) as well as increased in levels of anti-apoptotic genes *A1*, *BCL-2*, *BCL-X$_L$*, *BCL-W* and decreased in levels of pro-apoptotic genes *BID* and *BIM$_{EL}$* in ET and PMF. Furthermore, Tognon and colleagues (2012) found an association between JAK2 allele burden with *BAX*, *BIK* and *BAD* gene expression and correlation between JAK2 allele burden and *A1*, *BAX* and *BIK* gene expression. [45, 46]

Apoptosis deregulation is expected in MPN JAK2V617F positive diseases because the JAK2 pathway involves the participation of numerous genes in many vital cell processes including cell death and proliferation. However, the causes of the apoptosis alteration in JAK2V617F negative patients remain unclear.

Cell Cycle: Concepts and Alterations in MPN

The cell division process or cell cycle consists on the DNA replication and chromosome segregation between two new cells and it is divided basically in two steps denominated mitosis and interphase. The first one comprises the nuclear division and it has four stages: prophase, metaphase, anaphase and telophase. The second one, localized between two M phase, is split in general in three stages: G1, S and G2. G1 phase comprises the preparation for DNA replication, the S phase to the DNA synthesis and the G2 phase to preparation for mitosis. As an alternative, the cell can go to a steady state denominated G0 phase, after the G1 phase instead of going to S phase (reviewed by Vermeulen, 2003). [47]

All these process are very regulated by many proteins. The proteins Cyclin-Dependent Kinases (CDK) are key molecules involved in cell cycle regulation because they add phosphate to other key proteins as cyclins. For example, CDK4 and Cyclin D interaction is essential to G1 phase initiation while CDK2 and CYCLIN E interaction is essential for G1 to S phase progression. Another class of proteins, the CDK inhibitors, is also important members in the regulation process. These proteins are split into two families: the INK4 family (e.g.,, p15, p16, p18) and the Cip/Kip family (e.g.,, p21, p27, p57). Their expression is regulated by intrinsic and extrinsic factors and they are very important for the adult hematopoietic stem cell proliferation control. For example, TGF-b induces CDK inhibitors as p21, p27 and INK family while Notch signaling blocks this CDK inhibitors promoting proliferation. [44], [48] p21 and p27 are also involved in the control of the self-renewal of neural, intestinal and hematopoietic progenitors [48].

The suppressor of tumor p53, transcribed after DNA damage for stopping cell cycle, is an important regulator of p21, which contains in its gene promoter a region a p53-binding site. [49] As a substrate of CDK, the protein Rb, codified by the Retinoblastoma (*Rb*) gene, also has a crucial role in cell cycle regulation. When it is phosphorylated, transcription factors for the gene involved in the S phase are able to act. [50] Abnormalities in the Rb pathway are found in the majority of the tumors, usually with an increased Rb degradation. By the way, *P53* and *Rb* are called tumor suppressor genes and their inactivation leads to dysfunction of proteins involved in cell cycle and cell proliferation. Mutation in *P53* gene (*TP53*) is found in about 50% of tumors, however in hematologic neoplasms, it is less frequent. In these diseases, it is more common to verify the attenuation of p53 activity. [47] Regarding Rb expression in hematologic neoplasms, it has been described altered in acute myeloblastic and acute lymphoblastic leukemias, as well as other cell cycle related genes such as the *p16/INK4a* tumor suppressor gene, frequently silenced in acute myelogenous leukemia and several types of lymphoma. [51]

Another important cell cycle regulation mechanism is related to the control of the molecules localization inside the cell organelles. For example, the cyclins molecules receive signals to indicate if and when it should be transported to the nucleus and proteins kinases activators of CDK, as CdC27, are kept at the cytoplasm by interaction with other proteins as protein 14-3-3. [52] Cyclin-dependent kinase (CDK) and CDK inhibitors are frequently deregulated in neoplasms.

The cell cycle process has important check and restriction points. The checkpoints are placed before the S phase (G1-S checkpoint) and after replication (G2-M checkpoint) and the restriction point is a no-return point in G1. These points are necessary to avoid errors and mutation in the new cell's DNA. [48, 53]

In CML, the BCR-ABL provides a mitogenic signal that results in overexpression of CCND2 (D-type cyclins, which forms active complexes with CDK4 and CDK6) and facilitates the G1-S phase transition in leukemic cells. [54] Is known that p21 plays an essential role in growth arrest following DNA damage by binding and inhibiting cyclin/CDK complexes. This protein was originally considered a negative regulator of the cell cycle and a tumor suppressor. [55] Nowadays is believed that the functions of p21 depend on its intracellular localization. In the nucleus, it serves as a negative cell cycle regulator and tumor suppressor. When p21 is localized in the cytoplasm, it acts as an oncogene by protecting cells against apoptosis. [56] In addition, levels of p21 often determine the cellular response to various drugs that showed that when BCR-ABL positive cells were treated with antitumor drug, BCR-ABL-induced expression of p21 was found localized exclusively in the nucleus [57].

In BCR-ABL-positive cells, this study showed that after the treatment the p21 decreased cell proliferation; however, it did not change the level of spontaneous apoptosis suggesting that p21 acts as a negative regulator of the cell cycle and induces cell cycle arrest at the G0/G1 phase in leukemic cells treated with an antitumor drug.

Furthermore, another study showed that there are evidences that confer a protective advantage against apoptosis, which appears to be correlated with a cytoplasmic translocation of this protein. [58] Based on these findings, it is necessary to clarify the role of p21 and CDKN1A in BCR-ABL positive cells.

In Ph(-) MPN patients, Daí e Krantz (2001) studied the expression of cell cycle related genes in highly purified human erythroid colony-forming cells (ECFCs) from PV patients and they showed that *p16INK4a* and *p14ARF* gene expression was enhanced in 11 patients with PV, as well the p27 gene in PV ECFCs and Western blot analysis showed that the INK4a protein (p16INK4a) was increased in PV ECFCs. [59] They also investigated mutations in the *INK4a* or *ARF* gene in 10 patients with PV but they did not find one. *p16INK4a* expression in PV was dramatically increased without a significant change in ECFC cell cycle compared with normal ECFCs. They suggested that these findings might represent a cellular response to an abnormality of a downstream regulator of proliferation such as CYCLIN D, CDK4/CDK6, Rb, or E2F. In cellular models, changes in *p27* and *Src* genes were linked to JAK2 V617F mutation. (59) Furuhata et al (2009) analyzed cell cycle related molecules in JAK2V617F-Ba/F3 and mock-Ba/ F3. JAK2V617F-Ba/F3, but not mock-Ba/F3, which showed a deregulation of *p27Kip1*, the cell cycle regulator at the G1 to S transition, due to high Skp2 expression, a subunit of ubiquitin E3 ligase, through the STAT binding in the Skp2 promoter. Furthermore, constitutively active STAT5 or STAT3 induced aberrant p27Kip1 expression of Ba/F3 cells and in BCR/ABL-transfected Ba/F3. These authors suggested that their results indicate a regulatory mechanism by which JAK2V617F modulates *SKP2* gene expression through the STAT transcription factors. [60] More recently, Nakatake et al (2011) have demonstrated in the Ba/F3-EPOR cell line and HSC CD34[+] isolated from Ph(-) MPN patients that p53 degradation is augmented in these patients due to MDM2 accumulation promoted by the JAK2 mutated. [61]

Rodrigues, Reddy and Sattler (2008) have published a review regarding cell cycle regulation by TK in myeloid neoplasms considering oxidative stress influences in these diseases. They emphasized the participation of CYCLIN D proteins and p27Kip for S phase entry and the possibility of the p27Kip regulation downstream of STAT5, in combination with redox-dependent processes, may significantly contribute to the regulation of G1/S-phase transition. They discuss the fact that JAK2/STAT5 pathway activation would be sufficient for the induction of elevated levels of reactive oxygen species

(ROS) and CYCLIN D2, as well as the suppression of the cell cycle inhibitor p27Kip. [62]

Further studies regarding the cell cycle process regulation in Ph (-) MPN is still required for comprehension of their participation in the MPN physiopathology besides JAK/STAT pathway constitutive activation and apoptosis deregulation. In 2010, a study investigated the methylation profile of some cell cycle related gene (*CDKN2A, CDKN2B, XAF1, CDH13, JUNB*) in 31 bone marrow and 21 peripheral blood samples from ET patients but no evidence was found of hypermethylation. [63] Epigenetic regulation is a field into molecular studies, which has evolved rapidly in recent years. As discussed in the next section, the DNA methylation and histones modifications profile may explain many alterations in gene expression and could be a mechanism involved in MPN physiopathology.

Epigenetic: Concepts and Alterations in MPN

The term "epigenetic regulation" refers to DNA or DNA-associated proteins modifications, which do not change the DNA sequence but change the availability of the DNA strand to be reading and transcribed into RNA molecules. This gene expression control occurs, for example, by addition of methyl groups at the 5' position of cytosine nucleotides located in the CpG island at the promoters, centromeric or intergenic regions. [64, 65]

These modifications lead to chromatin compaction and transcription factors displacement as well as methyl binding proteins recruitment. Functionally, DNA methylation is associated with non-codifying regions of the genome. It shows a lineage-specific pattern established during the embrionary development. [65, 66] Besides, the post-translation modification histone tails, including phosphorylation, methylation, and ubiquitination, is also an important mechanism to control gene expression, being essential for the maintenance of eukaryotic cell function. Alterations in histone modifications due to deregulation of histones modifying protein activity have the potential to impact and drive the development of numerous human conditions and, in particular, cancer. [67] For example, deregulation function of histone acetyltransferases (HATs) and the histone deacetylases (HDACs) have been linked to many diseases and HDAC inhibitors (HDACIs) that show promising clinical utility in cancer treatment. The anti-tumor activity of HDACIs has been confirmed in a number of Phase I/II clinical trials. [68]

In CML, much evidence indicates that the epigenetic abnormalities, mainly different DNA methylation status at CpG sites, are involved in the pathogenesis of various types of human malignancies including the disease progression of CML. [69, 70] Recent studies have been performed to understand the role of aberrant methylation in the progression of CML.

Nguyen et al (2000) demonstrated that in the beginning of the course of CML there is an allele specific methylation in the translocated ABL1 promoter, a phenomenon that is unique to CML. Furthermore, an aberrant methylation status of p53 gene was described. BCR-ABL positive cells that harbor mutant p53 have shown abnormal methylation standards in the CpG sites, which alter the conformation of DNA and are also believed to play a role in genomic stability. This process appears to be a reciprocal action between p53 deregulation and changed methylation patterns with the progression of CML. [71]

According to Avramouli et al. (2009) a few studies have investigated the methylation status of individual tumor-suppressor genes in CML and these studies were limited by the relatively random choice of genes examined, which was based on studies of other malignancies. [72] However, a study showed that hypermethylation of the ATG16L2 gene promoter has been associated with a poor response to imatinib treatment. [73] In addition, Jelinek et al (2011) investigated the epigenetic impact of DNA methylation in CML. To clarify the role, these researchers analyzed 120 patients with CML for methylation of promoter-associated CpG islands of 10 genes. Five genes were identified by DNA methylation screening in the BCR-ABL positive cell line and 3 genes in patients with myeloproliferative neoplasms. The CDKN2B gene was selected for its frequent methylation in myeloid malignancies and ABL1 as the target of BCR-ABL translocation. The authors found that DNA methylation was strongly associated with disease progression and resistance to imatinib in CML [74].

Moreover, a study performed by Qian et al (2009) showed an abnormal methylation in the death-associated protein kinase 1 (DAPK1). The DAPK1 epigenetic changes have been recognized as a mechanism, which contributes to the development of CML. To elucidate the role of DAPK1 in CML, the methylation status of DAPK1 was evaluated in 49 patients with CML. The aberrant methylation of the DAPK1 gene was found in 25 of 49 (51.0%) cases of CML, not in all controls. No correlation was found between DAPK1 gene methylation and the age, hematologic parameters, chromosomal abnormalities, the types and levels of BCR/ABL transcripts of CML patients. However, correlation could be observed between the sex and the status of DAPK1 methylation in CML patients. Furthermore, there was a significant correlation between DAPK1 (methylation and the stages of CML. These results suggested that the DAPK1 promoter methylation plays an important role in the progression of CML [75].

In MPN Ph (-), after the description of JAK2V617F mutation, other mutations were described. For example, a mutation in the gene TET2 (TET family -"ten-eleven translocation") was shown in 12% of 320 MPN, LMA and MDS patients. The TET family members play an important role in epigenetic regulation.[76] As consequence of the TET2 mutation, a low level of 5-methylcytosine (5-mC) to 5-methylcytosine hydroxylation (5-hmC) conversion was observed. Recent studies describe that the balance between 5-mC/5-hmC are tightly associated to the balance between lineage commitment and pluripotency indicating the role of this gene in the epigenetic regulation, however it is not known if this mutation is an event which occurs before or after the JAK2V617F mutation [77].

Disturbs in epigenetic regulation in SOCS genes (suppressor cytokine of signaling protein system) were also described by many authors. [78-80] These authors suggested that SOCS 1 e 3 hypermethylation leads to JAK/STAT pathway activation and cytokine activating pathways amplification, participating in Ph(-) MPN physiopathology as a complementary mechanism to JAK2V617F mutation.

Capello and colleagues (2008) investigated epigenetic and genetic inactivation of SOCS3, SOCS1 and PTPN6 (negative regulators of the JAK-STAT pathway) in 112 CMPD and 20 acute myeloid leukemia (AML) post-CMPD. They described that SOCS3 methylation occurred with high frequency in both CMPD and AML post-CMPD and was associated with transcriptional silencing whereas SOCS1 and PTPN6 methylation was observed in only a fraction of CMPD and AML post-CMPD, indicating that methylation of SOCS3 and, to a lesser extent, SOCS1 and PTPN6 may act as an alternative or complementary mechanism to JAK2 mutations, enhancing cytokine signal transduction. [80] One year before, Jost et al (2007) analyzed the methylation status of 13 genes (SOCS1, SHP-1, E-Caderin, MGMT, TIMP-2, TIMP-3, p15, p16, p73, DAPK1, RASSF1A, RARbeta2 e hMLH1) in 39 MPN patients and they showed one hypermethylated gene in 15/39 patients with MPN and 6/39 samples with different methylation pattern of SOCS1 gene. [81] Two works published in 2013 investigated the methylation profile of several genes in Ph (-) MPN patients. One of them, performed by Nischal and colleagues, conducted an assay based on restriction enzymes for methylation analysis denominated HELP (HpaII tiny fragment enriched by LM-PCR) in PV, ET,

PMF and healthy controls samples. They concluded that PV and ET are characterized by aberrant promoter hypermethylation, whereas PMF is an epigenetically distinct subgroup characterized by both aberrant hyper- and hypomethylation.

According to them, in PV and ET patients, the altered methylation genes were involved mainly in cell signaling pathways or were linked to transcription factors expression. In PMF, the genes with aberrant methylation profile were involved in inflammatory pathways. They also found association between methylation status and the presence of mutations in ASXL1 gene in PMF and with TET2 mutations in MPN, but they did not find any correlation with the presence of JAK2V617F mutation. [82] The second study, published by Pérez et al, analyzed 71 MPN patients (24 PV, 23 ET and 24 PMF) using genome-wide DNA methylation arrays. In contrast to Nischal et al findings, this study describes that PV, ET and PMF patients showed a similar aberrant DNA methylation pattern when compared to control samples. Pérez et al described that MPN patients presented the genes from NF-κB pathway differentially methylated, indicating that they may be involved in the pathogenesis of these diseases. This work also analyzed 13 transformed MPN cases and detected an increased number of differentially methylated regions linked to chronic myeloproliferative neoplasms. [83]

As mentioned above, the histones deacetylases (HDACs) also have a critical role in modulating gene expression and they are involved in many cancers. Regarding Ph (-) MPN, it was described that the HDAC enzyme had enhanced activity in CD34$^+$ cells from patients with PMF and that these raised levels were correlated to the degree of splenomegaly. [84] Another indication of HDACs participation in the Ph (-) MPN pathophysiology were the results published by Skov et al [85].

They performed a global gene expression profiling of whole blood from patients with Ph (-) MPN and described a marked deregulation of HDAC genes, with the highest expression levels being found in patients with ET, PMF. They also observed that the HDAC6 gene is progressively expressed in patients with ET, PV and PMF, reflecting a steady accumulation of abnormally expressed HDAC6 during disease evolution. These results point out the HDACs relevance as epigenetic targets to consider the design of new treatments for MPN.

MicroRNA: Small Non-Coding RNA Regulating Gene Expression

MicroRNAs (miRNAs) are noncoding RNAs 18–25 nt in length that regulate many cellular functions including cell proliferation, differentiation, and apoptosis by silencing specific target genes through translational repression or direct mRNA degradation. [86, 87] miRNAs are well conserved during evolution, and it has been estimated that about 250–600 miRNAs have been evolutionarily conserved in vertebrates. Although the detailed functions of the growing number of miRNAs identified in the mammalian genome are far from being completely characterized, studies have indicated that deregulated expression of specific miRNAs have been associated with diseases including solid tumors (lung, breast, colorectal cancer among others) and hematological malignancies (MPN, chronic lymphocytic leukemia, acute promyelocytic leukemia, acute lymphocytic leukemia and chronic myeloid leukemia) [44]

Distinct miRNA expression signatures are known to participate in regulatory pathways involved in the development and progression of many types of tumors. Since miRNAs affect a wide range of molecular pathways, miRNA expression studies in human cancer may provide a better understanding of the molecular pathways contributing to cancer pathogenesis and their progression [88].

Most studies about miRNA expression in CML explore the expression of specific miRNAs. [89, 90] Cimmino et al. (2005) demonstrated that miR-15a, and miR-16-1 negatively regulates the anti-apoptotic gene *BCL-2*, whose expression is elevated in lymphomas and leukemias. [91] Thus, it was demonstrated that deletion of miR-15a, and miR-16-1 results in increased expression of the *BCL-2* anti-apoptotic gene, which may promote leukemic cells resistance to death and potentially contributes to chronic lymphoid leukemia (CLL) leukemic phenotype. [92, 93].

In CML, the abnormal expression of miRNAs miR-10a, miR-196, miR-150a, miR-151, miR-451, miR-cluster miR34a and 17-92 have been described and are associated with leukemogenesis and BCR-ABL oncoprotein TK leading to cell cycle deregulation and apoptosis resistance. [94]

Increased expression of the cluster miR-17-92 was observed in patients in the chronic phase of CML, compared to patients in blast crisis. [94] Also in CML, it was shown that the low expression of miR-10a results in abnormal cell proliferation by disrupting gene expression USF2. [95]

Lopotová et al (2011) observed a correlation between the expression of miR-451 and increased TK activity of the BCR-ABL. This work was undertaken in newly diagnosed patients, expression of miR-451 was decreased, whereas patients treated with imatinib presented elevated levels of miR-451. [96]

More recently, Rokah et al (2012) detected miRNAs whose expression is downregulated in CML. In this study 3 miRNAs (miR-31, miR-155, and miR-564) were identified to be abnormally downregulated in CML cell lines and in patients with CML. [97] These data suggest that the low expression of these miRNAs is dependent on BCR-ABL TK activity as demonstrated by their increased expression in leukocytes of an imatinib treated patient.

Bueno et al. (2008) investigated if the expression of miR-203 is also downregulated by epigenetic mechanisms. [98] The results showed that the remaining copy of the DNA, miR-203 is muted by the loss of one allele and also by CpG hypermethylation of the promoter region. Moreover, this group identified ABL1 as a target of miR-203. The data obtained in this study indicated that both genetic and epigenetic mechanisms involved the silencing of miR-203, increases the ABL1 and BCR-ABL1 oncogene in hematologic malignancies.

Therefore, miRNAs and altered epigenetic mechanisms seem to play a key role in the pathophysiology of hematologic malignancies, such as CML.

According to these studies, further studies are required to address the molecular mechanism by which BCR-ABL is responsible for the downregulated expression of these miRNAs and whether this downregulation is essential for the transforming role of BCR-ABL in CML.

From 2007, many groups started working on miRNA research in myeloid cells and Ph(-) MPN. Three expression patterns of miRNAs in normal erythropoiesis comprising a progressive downregulation of miR-150, miR-155, miR-221, miR-222; upregulation of miR-451, miR-16 at late stages of erythropoiesis; and biphasic regulation of miR-339, miR-378 were observed and miR-451 was suggested as miR erythroid-specific. [99] Guglielmelli et al. (2007) described up-regulated expression of miRNA-182 and -183 in patients with PMF, which was correlated to JAK2 allele burden.

In 2008, O'Connel et al reported the relationship between a miR-155 and granulocyte/monocyte (GM) expansion with pathological features characteristic of myeloid neoplasms and Bruchova et al. (2008) described that the abnormal expression of miR-let-7, miR-451 and miR-130 was associated with down-regulation of target genes expression, which are important for red cell differentiation and proliferation. [100, 101] In 2009 and 2010, more studies were published associating microRNAs and Ph(-) MPN. Hussein et al (2009) showed that miRNA 17-5p, 20a and 126 are constitutively are expressed in Ph (-) MPN megakaryocytopoiesis while low or absent miRNA 10a seem to be correlated with strong megakaryocytic HOXA1 protein expression. [102] miR-28 was also found overexpressed in platelets of a fraction of MPN patients, while it was expressed at constant low levels in platelets from healthy subjects. It was showed that the autonomous growth of hematopoietic cell lines induced by constitutive activation of STAT5 was associated with increased miR-28 expression [103].

miR-16, a miRNA differentially expressed in other hematologic diseases, was found abnormally increased in PV CD34$^+$ cells as a consequence of preferential expression of miR-16-2 on chromosome 3 rather than of miR-16-1 on chromosome 13. [104] Furthermore, it was shown that forced expression of miRNA-16 in normal CD34$^+$ cells stimulated erythroid cell proliferation and maturation while exposure of PV CD34$^+$ cells to small interfering RNA against pre-miR-16-2 reduced erythroid colonies and largely prevented formation of erythropoietin-independent colonies; and myeloid progenitors remained unaffected. The same authors performing experiments with JAK2 knock down cell lines showed that miR-16 was independent of JAK/STAT pathway activation during abnormal erythropoiesis [104].

Since the literature showed the involvement of microRNA in the apoptosis process, and there is plenty of evidence of the deregulated apoptosis process in MPN, Nunes et al (2013) quantified by real-time PCR the expression of several apoptomiRs in CD34$^+$ and leukocytes of Ph(-) MPN patients. Nunes et al showed an abnormal expression of miR-26a, -130b, -21, -29c, -let-7d, -15a and -16 apoptomiRs in patients with PV, ET and PMF, indicating the possibility of using mRNAs as new therapeutic targets or biomarkers for MPN. [105].

High throughput techniques allowed a wide investigation of gene expression and recently Lin et al (2013) found that 61 miRNAs were significantly deregulated in CD34+ cells from Ph(-) MPN patients compared with controls. (106) Using this global approach, they verified different miRNA expression profiles between PV (JAK2 V617F) and ET (JAK2 wild type) patients. However, they also observed that the expression of miR-134, -214 and -433 were not affected by changes in JAK2 activity, suggesting that additional signaling pathways could be responsible for the deregulation of these miRNAs in MPN [106].

Also in 2013, Zhan et al. (2013) performed miRNA expression profiling by oligonucleotide microarray analysis in purified peripheral blood CD34$^+$

cells from PV and healthy donors and they concluded that deregulated miRNAs could represent an important mechanism by which the PV erythrocytosis and ET thrombocytosis phenotypes were determined [107].

Regarding the possibility of using microRNA as a molecular marker, Gebauer et al. (2013) evaluated by quantitative RT-PCR in PV, ET, early PMF and normal hematopoiesis in the differential expression of miRNAs previously described as associated with myelopoiesis and myeloproliferative pathogenesis. They found an aberrant expression of miRNAs 10a in ET and PMF and mir-150 in PV and PMF. They also found a correlation between miR-150 and both JAK2 allele burden and peripheral blood counts, concluding that these miRNAs could be a potential marker for oncomiRs in the differential diagnosis of Ph (-) MPN [108].

In conclusion, the findings about miRNAs indicate their participation in MPN physiopathology, contributing to the accumulation of myeloid cells. More studies are necessary to conclude the link between miRNA and the constitutively activated TK and the patients' clinical presentations, as well as their use as molecular markers. Table 1 draws attention to many microRNAs deregulated expression in MPN.

TAKE HOME MESSAGES

- Myeloproliferative neoplasms (MPN) are hematological diseases characterized by myeloproliferation/myeloaccumulation of hematopoietic mature cells without a specific stimulus;
- Chronic Myeloid Leukemia (CML), Polycythemia Vera (PV), Essential Thrombocythemia (ET) and Primary Myelofibrosis (PMF) present cell resistance to apoptosis and cycle cell deregulation;
- In MPN, the cells resistance to apoptosis and cell cycle deregulation seem to be linked to BCR-ABL and JAK2 tyrosine kinase constitutive activation;
- In CML, several evidences indicate that epigenetic abnormalities, mainly altered DNA methylation status, are involved in its pathogenesis;
- Mutations in other genes besides JAK2, such as TET2 (TET family - "ten-eleven translocation"), were also described in MPN negative for Ph;

- Abnormal miRNA expression signatures are known to participate in regulatory pathways involved in leukemogenesis and progression of many hematological diseases, as well as MPN.

Table 1. The microRNAs deregulated expression in MPN

Hematologic disease	miRNAs	Relation to Disease Pathogenesis	References
CML (Ph+)	miR-15a and miR-16-1	Negatively regulates the anti-apoptotic gene *BCL-2*	[91, 92, 93]
	miR-10a, miR-196, miR-150a, miR-151, miR-451, miR-cluster miR34a and 17-92, miR-31, miR-155, miR-564 and miR-203	Downregulated expression in CML is related to increased BCR-ABL TK activity	[94, 95,96, 98]
MPN (Ph⁻)	miRNA-182 and – miR-183	Abnormal expression in patients with PMF, which is correlated to JAK2 allele burden	[104]
	miR-let-7, miR-451 and miR-130	Down-regulation is correlated with alterations in red cell differentiation and proliferation	[100, 101]
	miR-28	Overexpressed in platelets of some Ph (-) MPN patients leading to the constitutive activation of STAT5	[103]
	miR-16	Abnormal expression in patients with PV, ET and PMF	[104]
	miR-26a, miR-130b, miR-21, miR-29c, miR-let-7d, miR-15a and miR-16	Abnormal expression in patients with PV, ET and PMF linked to apoptosis resistance	[105]
	miR-10a and miR-150	Aberrant myelopoiesis and cell blood count in ET and PMF as well as for PV and PMF, respectively	[108]

REFERENCES

[1]   Vardiman JW, Thiele J, Arber DA, Brunning RD, Borowitz MJ, Porwit A, et al. The 2008 revision of the World Health Organization (WHO) classification of myeloid neoplasms and acute leukemia: rationale and important changes. *Blood.* 2009;114(5):937-51.

[2]   Melo JV, Barnes DJ. Chronic myeloid leukaemia as a model of disease evolution in human cancer. *Nat. Rev. Cancer.* 2007;7(6):441-53.

[3]   Cortez D, Kadlec L, Pendergast AM. Structural and signaling requirements for BCR-ABL-mediated transformation and inhibition of apoptosis. *Mol. Cell. Biol.* 1995;15(10):5531-41.

[4]   Radich JP. Chronic myeloid leukemia 2010: where are we now and where can we go? *Hematology Am. Soc. Hematol. Educ. Program.* 2010;2010:122-8.

[5]   Perrotti D, Jamieson C, Goldman J, Skorski T. Chronic myeloid leukemia: mechanisms of blastic transformation. *J. Clin. Invest.* 2010;120(7):2254-64.

[6]   Faderl S, Kantarjian HM, Talpaz M. *Chronic myelogenous leukemia: update on biology and treatment. Oncology* (Williston Park). 1999;13(2):169-80; discussion 81, 84.

[7]   Hehlmann R, Hochhaus A, Baccarani M, LeukemiaNet E. Chronic myeloid leukaemia. *Lancet.* 2007;370(9584):342-50.

[8]   Heaney NB, Holyoake TL. Therapeutic targets in chronic myeloid leukaemia. *Hematol Oncol.* 2007;25(2):66-75.

[9]   Cotta CV, Bueso-Ramos CE. New insights into the pathobiology and treatment of chronic myelogenous leukemia. *Ann. Diagn. Pathol.* 2007;11(1):68-78.

[10] Deininger MW, Vieira S, Mendiola R, Schultheis B, Goldman JM, Melo JV. BCR-ABL tyrosine kinase activity regulates the expression of multiple genes implicated in the pathogenesis of chronic myeloid leukemia. *Cancer Res.* 2000;60(7):2049-55.

[11] Cross NC. Genetic and epigenetic complexity in myeloproliferative neoplasms. *Hematology Am. Soc. Hematol Educ. Program.* 2011;2011:208-14.

[12] Mehta J, Wang H, Iqbal SU, Mesa R. Epidemiology of Myeloproliferative neoplasms (MPN) in the United States. *Leuk Lymphoma.* 2013.

[13] Scherber R, Dueck AC, Johansson P, Barbui T, Barosi G, Vannucchi AM, et al. The Myeloproliferative Neoplasm Symptom Assessment

Form (MPN-SAF): international prospective validation and reliability trial in 402 patients. *Blood*. 2011;118(2):401-8.

[14] Mitelman F. The cytogenetic scenario of chronic myeloid leukemia. *Leuk Lymphoma*. 1993;11 Suppl 1:11-5.

[15] Nowell PC, Hungerford Da. Chromosome studies on normal and leukemic human leukocytes. *J. Natl. Cancer Inst*. 1960;25:85-109.

[16] Deininger MW, Goldman JM, Melo JV. The molecular biology of chronic myeloid leukemia. *Blood*. 2000;96(10):3343-56.

[17] Hamdane M, David-Cordonnier MH, D'Halluin JC. Activation of p65 NF-kappaB protein by p210BCR-ABL in a myeloid cell line (P210BCR-ABL activates p65 NF-kappaB). *Oncogene*. 1997;15(19):2267-75.

[18] von Bubnoff N, Duyster J. Chronic myelogenous leukemia: treatment and monitoring. *Dtsch Arztebl Int*. 2010;107(7):114-21.

[19] Johansson B, Fioretos T, Mitelman F. Cytogenetic and molecular genetic evolution of chronic myeloid leukemia. *Acta Haematol*. 2002;107(2): 76-94.

[20] James C, Ugo V, Le Couédic JP, Staerk J, Delhommeau F, Lacout C, et al. A unique clonal JAK2 mutation leading to constitutive signalling causes polycythaemia vera. *Nature*. 2005;434(7037):1144-8.

[21] Baxter EJ, Scott LM, Campbell PJ, East C, Fourouclas N, Swanton S, et al. Acquired mutation of the tyrosine kinase JAK2 in human myeloproliferative disorders. *Lancet*. 2005;365(9464):1054-61.

[22] Levine RL, Wadleigh M, Cools J, Ebert BL, Wernig G, Huntly BJ, et al. Activating mutation in the tyrosine kinase JAK2 in polycythemia vera, essential thrombocythemia, and myeloid metaplasia with myelofibrosis. *Cancer Cell*. 2005;7(4):387-97.

[23] Kralovics R, Passamonti F, Buser AS, Teo SS, Tiedt R, Passweg JR, et al. A gain-of-function mutation of JAK2 in myeloproliferative disorders. *N Engl J Med*. 2005;352(17):1779-90.

[24] Zhao R, Xing S, Li Z, Fu X, Li Q, Krantz SB, et al. Identification of an acquired JAK2 mutation in polycythemia vera. *J. Biol. Chem*. 2005;280(24):22788-92.

[25] Vainchenker W, Delhommeau F, Villeval J. Molecular pathogenesis of the myeloproliferative diseases. Hematology education: the education program for the annual congress of the European Hematology Association; Berlin, Germany: European Hematology Association; 2007. p. 239-46.

[26] Speletas M, Katodritou E, Daiou C, Mandala E, Papadakis E, Kioumi A, et al. Correlations of JAK2-V617F mutation with clinical and laboratory

findings in patients with myeloproliferative disorders. *Leuk Res.* 2007;31(8):1053-62.

[27] Reilly JT. Idiopathic myelofibrosis: pathogenesis to treatment. *Hematol. Oncol.* 2006;24(2):56-63.

[28] Delhommeau F, Jeziorowska D, Marzac C, Casadevall N. Molecular aspects of myeloproliferative neoplasms. *Int. J. Hematol.* 2010;91(2):165-73.

[29] Vainchenker W, Delhommeau F, Constantinescu SN, Bernard OA. New mutations and pathogenesis of myeloproliferative neoplasms. *Blood.* 2011;118(7):1723-35.

[30] Pietra D, Brisci A, Rumi E, Boggi S, Elena C, Pietrelli A, et al. Deep sequencing reveals double mutations in cis of MPL exon 10 in myeloproliferative neoplasms. *Haematologica.* 2011;96(4):607-11.

[31] Pardanani A, Lasho T, Finke C, Oh ST, Gotlib J, Tefferi A. LNK mutation studies in blast-phase myeloproliferative neoplasms, and in chronic-phase disease with TET2, IDH, JAK2 or MPL mutations. *Leukemia.* 2010;24(10):1713-8.

[32] Vainchenker W, Delhommeau F, Constantinescu SN, Bernard OA. New mutations and pathogenesis of myeloproliferative neoplasms. *Blood.* 2011.

[33] Grand FH, Hidalgo-Curtis CE, Ernst T, Zoi K, Zoi C, McGuire C, et al. Frequent CBL mutations associated with 11q acquired uniparental disomy in myeloproliferative neoplasms. *Blood.* 2009;113(24):6182-92.

[34] Rathmell JC, Thompson CB. Pathways of apoptosis in lymphocyte development, homeostasis, and disease. *Cell.* 2002;109 Suppl:S97-107.

[35] Jin Z, El-Deiry WS. Overview of cell death signaling pathways. *Cancer Biol Ther.* 2005;4(2):139-63.

[36] Amarante-Mendes GP. The regulation of apoptotic cell death. *Braz. J. Med. Biol. Res;* 1999. p. 1053-61.

[37] *Borner C. The Bcl-2* protein family: sensors and checkpoints for life-or-death decisions. *Mol. Immunol.* 2003;39(11):615-47.

[38] Irmler M, Thome M, Hahne M, Schneider P, Hofmann K, Steiner V, et al. Inhibition of death receptor signals by cellular FLIP. *Nature.* 1997;388(6638):190-5.

[39] Brumatti G, Weinlich R, Chehab CF, Yon M, Amarante-Mendes GP. Comparison of the anti-apoptotic effects of Bcr-Abl, Bcl-2 and Bcl-x(L) following diverse apoptogenic stimuli. *FEBS Lett.* 2003;541(1-3):57-63.

[40] Gutiérrez-Castellanos S, Cruz M, Rabelo L, Godínez R, Reyes-Maldonado E, Riebeling-Navarro C. Differences in BCL-X(L)

expression and STAT5 phosphorylation in chronic myeloid leukaemia patients. *Eur. J. Haematol.* 2004;72(4):231-8.

[41] Keeshan K, Mills KI, Cotter TG, McKenna SL. Elevated Bcr-Abl expression levels are sufficient for a haematopoietic cell line to acquire a drug-resistant phenotype. *Leukemia.* 2001;15(12):1823-33.

[42] Kuribara R, Honda H, Matsui H, Shinjyo T, Inukai T, Sugita K, et al. Roles of Bim in apoptosis of normal and Bcr-Abl-expressing hematopoietic progenitors. *Mol. Cell. Biol.* 2004;24(14):6172-83.

[43] Aichberger KJ, Mayerhofer M, Krauth MT, Skvara H, Florian S, Sonneck K, et al. Identification of mcl-1 as a BCR/ABL-dependent target in chronic myeloid leukemia (CML): evidence for cooperative antileukemic effects of imatinib and mcl-1 antisense oligonucleotides. *Blood.* 2005;105(8):3303-11.

[44] Zhang L, Zhao H, Sun A, Lu S, Liu B, Tang F, et al. Early down-regulation of Bcl-xL expression during megakaryocytic differentiation of thrombopoietin-induced CD34+ bone marrow cells in essential thrombocythemia. *Haematologica.* 2004;89(10):1199-206.

[45] Tognon R, Gasparotto EP, Leroy JM, Oliveira GL, Neves RP, Carrara ReC, et al. Differential expression of apoptosis-related genes from death receptor pathway in chronic myeloproliferative diseases. *J. Clin. Pathol.* 2011;64(1):75-82.

[46] Tognon R, Gasparotto EP, Neves RP, Nunes NS, Ferreira AF, Palma PV, et al. Deregulation of apoptosis-related genes is associated with PRV1 overexpression and JAK2 V617F allele burden in Essential Thrombocythemia and Myelofibrosis. *J. Hematol. Oncol.* 2012;5:2.

[47] Vermeulen K, Van Bockstaele DR, Berneman ZN. The cell cycle: a review of regulation, deregulation and therapeutic targets in cancer. *Cell. Prolif.* 2003;36(3):131-49.

[48] Malumbres M, Barbacid M. Cell cycle, CDKs and cancer: a changing paradigm. *Nat Rev Cancer.* 2009;9(3):153-66.

[49] el-Deiry WS, Tokino T, Velculescu VE, Levy DB, Parsons R, Trent JM, et al. WAF1, a potential mediator of p53 tumor suppression. *Cell.* 1993;75(4):817-25.

[50] Weinberg RA. The retinoblastoma protein and cell cycle control. *Cell.* 1995;81(3):323-30.

[51] Spike BT, Macleod KF. The Rb tumor suppressor in stress responses and hematopoietic homeostasis. *Cell Cycle.* 2005;4(1):42-5.

[52] Peng CY, Graves PR, Thoma RS, Wu Z, Shaw AS, Piwnica-Worms H. Mitotic and G2 checkpoint control: regulation of 14-3-3 protein binding

by phosphorylation of Cdc25C on serine-216. *Science.* 1997;277(5331):1501-5.

[53] Kastan MB, Bartek J. Cell-cycle checkpoints and cancer. *Nature.* 2004;432(7015):316-23.

[54] Ando K, Ajchenbaum-Cymbalista F, Griffin JD. Regulation of G1/S transition by cyclins D2 and D3 in hematopoietic cells. *Proc. Natl. Acad. Sci. USA.* 1993;90(20):9571-5.

[55] Gartel AL, Tyner AL. The role of the cyclin-dependent kinase inhibitor p21 in apoptosis. *Mol. Cancer Ther.* 2002;1(8):639-49.

[56] Romanov VS, Pospelov VA, Pospelova TV. Cyclin-dependent kinase inhibitor p21(Waf1): contemporary view on its role in senescence and oncogenesis. *Biochemistry* (Mosc). 2012;77(6):575-84.

[57] Wang Y, Chen J, Wang L, Huang Y, Leng Y, Wang G. Fangchinoline induces G0/G1 arrest by modulating the expression of CDKN1A and CCND2 in K562 human chronic myelogenous leukemia cells. *Exp. Ther. Med.* 2013;5(4):1105-12.

[58] Coqueret O. New roles for p21 and p27 cell-cycle inhibitors: a function for each cell compartment? *Trends Cell Biol.* 2003;13(2):65-70.

[59] Dai C, Krantz SB. Increased expression of the INK4a/ARF locus in polycythemia vera. *Blood.* 2001;97(11):3424-32.

[60] Furuhata A, Kimura A, Shide K, Shimoda K, Murakami M, Ito H, et al. p27 deregulation by Skp2 overexpression induced by the JAK2V617 mutation. *Biochem. Biophys. Res. Commun.* 2009;383(4):411-6.

[61] Nakatake M, Monte-Mor B, Debili N, Casadevall N, Ribrag V, Solary E, et al. JAK2(V617F) negatively regulates p53 stabilization by enhancing MDM2 via La expression in myeloproliferative neoplasms. *Oncogene.* 2011;31(10):1323-33.

[62] Rodrigues MS, Reddy MM, Sattler M. Cell cycle regulation by oncogenic tyrosine kinases in myeloid neoplasias: from molecular redox mechanisms to health implications. *Antioxid Redox Signal.* 2008;10(10):1813-48.

[63] Chim CS, Fung TK, Liang R. Methylation of cyclin-dependent kinase inhibitors, XAF1, JUNB, CDH13 and soluble Wnt inhibitors in essential thrombocythaemia. *J. Clin. Pathol.* 2010;63(6):518-21.

[64] Lister R, Pelizzola M, Dowen RH, Hawkins RD, Hon G, Tonti-Filippini J, et al. Human DNA methylomes at base resolution show widespread epigenomic differences. *Nature.* 2009;462(7271):315-22.

[65] Rivera RM, Bennett LB. Epigenetics in humans: an overview. *Curr. Opin Endocrinol Diabetes Obes.* 2010;17(6):493-9.

[66] Mohtat D, Susztak K. Fine tuning gene expression: the epigenome. *Semin. Nephrol.* 2010;30(5):468-76.

[67] Thompson LL, Guppy BJ, Sawchuk L, Davie JR, McManus KJ. Regulation of chromatin structure via histone post-translational modification and the link to carcinogenesis. *Cancer Metastasis Rev.* 2013.

[68] Guo SQ, Zhang YZ. Histone deacetylase inhibition: an important mechanism in the treatment of lymphoma. *Cancer Biol. Med.* 2012;9(2):85-9.

[69] Roman-Gomez J, Castillejo JA, Jimenez A, Cervantes F, Boque C, Hermosin L, et al. Cadherin-13, a mediator of calcium-dependent cell-cell adhesion, is silenced by methylation in chronic myeloid leukemia and correlates with pretreatment risk profile and cytogenetic response to interferon alfa. *J. Clin. Oncol.* 2003;21(8):1472-9.

[70] Strathdee G, Holyoake TL, Sim A, Parker A, Oscier DG, Melo JV, et al. Inactivation of HOXA genes by hypermethylation in myeloid and lymphoid malignancy is frequent and associated with poor prognosis. *Clin. Cancer Res.* 2007;13(17):5048-55.

[71] Guinn BA, Mills KI. p53 mutations, methylation and genomic instability in the progression of chronic myeloid leukaemia. *Leuk. Lymphoma.* 1997;26(3-4):211-26.

[72] Avramouli A, Tsochas S, Mandala E, Katodritou E, Ioannou M, Ritis K, et al. Methylation status of RASSF1A in patients with chronic myeloid leukemia. *Leuk. Res.* 2009;33(8):1130-2.

[73] Dunwell T, Hesson L, Rauch TA, Wang L, Clark RE, Dallol A, et al. A genome-wide screen identifies frequently methylated genes in haematological and epithelial cancers. *Mol. Cancer.* 2010;9:44.

[74] Jelinek J, Gharibyan V, Estecio MR, Kondo K, He R, Chung W, et al. Aberrant DNA methylation is associated with disease progression, resistance to imatinib and shortened survival in chronic myelogenous leukemia. *PLoS One.* 2011;6(7):e22110.

[75] Qian J, Wang YL, Lin J, Yao DM, Xu WR, Wu CY. Aberrant methylation of the death-associated protein kinase 1 (DAPK1) CpG island in chronic myeloid leukemia. *Eur. J. Haematol.* 2009;82(2):119-23.

[76] Delhommeau F, Dupont S, Della Valle V, James C, Trannoy S, Massé A, et al. Mutation in TET2 in myeloid cancers. *N. Engl. J. Med.* 2009;360(22):2289-301.

[77] Odenike O, Thirman MJ, Artz AS, Godley LA, Larson RA, Stock W. Gene mutations, epigenetic dysregulation, and personalized therapy in myeloid neoplasia: are we there yet? *Semin Oncol.* 2011;38(2):196-214.

[78] Quentmeier H, Geffers R, Jost E, Macleod RA, Nagel S, Röhrs S, et al. SOCS2: inhibitor of JAK2V617F-mediated signal transduction. *Leukemia.* 2008;22(12):2169-75.

[79] Teofili L, Martini M, Cenci T, Guidi F, Torti L, Giona F, et al. Epigenetic alteration of SOCS family members is a possible pathogenetic mechanism in JAK2 wild type myeloproliferative diseases. *Int. J. Cancer.* 2008;123(7):1586-92.

[80] Capello D, Deambrogi C, Rossi D, Lischetti T, Piranda D, Cerri M, et al. Epigenetic inactivation of suppressors of cytokine signalling in Philadelphia-negative chronic myeloproliferative disorders. *Br. J. Haematol.* 2008;141(4):504-11.

[81] Jost E, do O N, Dahl E, Maintz CE, Jousten P, Habets L, et al. Epigenetic alterations complement mutation of JAK2 tyrosine kinase in patients with BCR/ABL-negative myeloproliferative disorders. *Leukemia.* 2007;21(3):505-10.

[82] Nischal S, Bhattacharyya S, Christopeit M, Yu Y, Zhou L, Bhagat TD, et al. Methylome profiling reveals distinct alterations in phenotypic and mutational subgroups of myeloproliferative neoplasms. *Cancer Res.* 2013;73(3):1076-85.

[83] Pérez C, Pascual M, Martin-Subero JI, Bellosillo B, Segura V, Delabesse E, et al. Aberrant DNA methylation profile of chronic and transformed classicphiladelphia-negative myeloproliferative neoplasms. *Haematologica.* 2013.

[84] Wang JC, Chen C, Dumlao T, Naik S, Chang T, Xiao YY, et al. Enhanced histone deacetylase enzyme activity in primary myelofibrosis. *Leuk Lymphoma.* 2008;49(12):2321-7.

[85] Skov V, Larsen TS, Thomassen M, Riley CH, Jensen MK, Bjerrum OW, et al. Increased gene expression of histone deacetylases in patients with Philadelphia-negative chronic myeloproliferative neoplasms. *Leuk Lymphoma.* 2012;53(1):123-9.

[86] Ambros V. The functions of animal microRNAs. *Nature.* 2004;431(7006):350-5.

[87] Zhao H, Wang D, Du W, Gu D, Yang R. MicroRNA and leukemia: tiny molecule, great function. *Crit .Rev. Oncol. Hematol.* 2010;74(3):149-55.

[88] Machova Polakova K, Koblihova J, Stopka T. Role of epigenetics in chronic myeloid leukemia. *Curr Hematol Malig Rep.* 2013;8(1):28-36.

[89] Borze I, Guled M, Musse S, Raunio A, Elonen E, Saarinen-Pihkala U, et al. MicroRNA microarrays on archive bone marrow core biopsies of leukemias--method validation. *Leuk. Res.* 2011;35(2):188-95.

[90] Hussein K, Büsche G, Muth M, Göhring G, Kreipe H, Bock O. Expression of myelopoiesis-associated microRNA in bone marrow cells of atypical chronic myeloid leukaemia and chronic myelomonocytic leukaemia. *Ann. Hematol.* 2011;90(3):307-13.

[91] Cimmino A, Calin GA, Fabbri M, Iorio MV, Ferracin M, Shimizu M, et al. miR-15 and miR-16 induce apoptosis by targeting BCL2. *Proc. Natl. Acad. Sci. USA.* 2005;102(39):13944-9.

[92] Esquela-Kerscher A, Slack FJ. Oncomirs - microRNAs with a role in cancer. *Nat. Rev. Cancer.* 2006;6(4):259-69.

[93] Zhang L, Huang J, Yang N, Greshock J, Megraw MS, Giannakakis A, et al. microRNAs exhibit high frequency genomic alterations in human cancer. *Proc. Natl. Acad. Sci. USA.* 2006;103(24):9136-41.

[94] Venturini L, Battmer K, Castoldi M, Schultheis B, Hochhaus A, Muckenthaler MU, et al. Expression of the miR-17-92 polycistron in chronic myeloid leukemia (CML) CD34+ cells. *Blood.* 2007;109(10):4399-405.

[95] Agirre X, Jiménez-Velasco A, San José-Enériz E, Garate L, Bandrés E, Cordeu L, et al. Down-regulation of hsa-miR-10a in chronic myeloid leukemia CD34+ cells increases USF2-mediated cell growth. *Mol. Cancer Res.* 2008;6(12):1830-40.

[96] Lopotová T, Záčková M, Klamová H, Moravcová J. MicroRNA-451 in chronic myeloid leukemia: miR-451-BCR-ABL regulatory loop? Leuk. Res. 2011;35(7):974-7.

[97] Rokah OH, Granot G, Ovcharenko A, Modai S, Pasmanik-Chor M, Toren A, et al. Downregulation of miR-31, miR-155, and miR-564 in chronic myeloid leukemia cells. *PLoS One.* 2012;7(4):e35501.

[98] Bueno MJ, Pérez de Castro I, Gómez de Cedrón M, Santos J, Calin GA, Cigudosa JC, et al. Genetic and epigenetic silencing of microRNA-203 enhances ABL1 and BCR-ABL1 oncogene expression. *Cancer Cell.* 2008;13(6):496-506.

[99] Bruchova H, Yoon D, Agarwal AM, Mendell J, Prchal JT. Regulated expression of microRNAs in normal and polycythemia vera erythropoiesis. *Exp. Hematol.* 2007;35(11):1657-67.

[100] O'Connell RM, Rao DS, Chaudhuri AA, Boldin MP, Taganov KD, Nicoll J, et al. Sustained expression of microRNA-155 in hematopoietic

stem cells causes a myeloproliferative disorder. *J. Exp. Med.* 2008;205(3):585-94.

[101] Bruchova H, Merkerova M, Prchal JT. Aberrant expression of microRNA in polycythemia vera. *Haematologica.* 2008;93(7):1009-16.

[102] Hussein K, Dralle W, Theophile K, Kreipe H, Bock O. Megakaryocytic expression of miRNA 10a, 17-5p, 20a and 126 in Philadelphia chromosome-negative myeloproliferative neoplasm. *Ann. Hematol.* 2009;88(4):325-32.

[103] Girardot M, Pecquet C, Boukour S, Knoops L, Ferrant A, Vainchenker W, et al. miR-28 is a thrombopoietin receptor targeting microRNA detected in a fraction of myeloproliferative neoplasm patient platelets. *Blood.* 2010;116(3):437-45.

[104] Guglielmelli P, Tozzi L, Bogani C, Iacobucci I, Ponziani V, Martinelli G, et al. Overexpression of microRNA-16-2 contributes to the abnormal erythropoiesis in polycythemia vera. *Blood.* 2011;117(25):6923-7.

[105] Nunes NS, Tognon R, Moura LG, Kashima S, Covas DT, Santana M, et al. Differential expression of apoptomiRs in myeloproliferative neoplasms. *Leuk. Lymphoma.* 2013.

[106] Lin X, Rice KL, Buzzai M, Hexner E, Costa FF, Kilpivaara O, et al. miR-433 is aberrantly expressed in myeloproliferative neoplasms and suppresses hematopoietic cell growth and differentiation. *Leukemia.* 2013;27(2):344-52.

[107] Zhan H, Cardozo C, Yu W, Wang A, Moliterno AR, Dang CV, et al. MicroRNA deregulation in polycythemia vera and essential thrombocythemia patients. *Blood Cells Mol. Dis.* 2013;50(3):190-5.

[108] Gebauer N, Bernard V, Gebauer W, Feller AC, Merz H. MicroRNA Expression and JAK2 Allele Burden in Bone Marrow Trephine Biopsies of Polycythemia Vera, Essential Thrombocythemia and Early Primary Myelofibrosis. *Acta. Haematol.* 2013;129(4):251-6.

In: Myeloid Cells
Editor: Spencer A. Douglas

ISBN: 978-1-62948-046-6
© 2013 Nova Science Publishers, Inc.

Chapter 4

USE OF ANIMAL MODELS TO EVALUATE MYELOID CELL DYSFUNCTION IN CANCER

Tracey L. Papenfuss and Julie Baker*
The Ohio State University, Columbus, Ohio, US

ABSTRACT

In recent years, the concept of using the body's own immune system to target and kill tumor cells has been an appealing approach to treat cancer. Cancer immunotherapy holds particular promise for treating metastatic disease but has met with relatively limited success due to our limited understanding of how the immune system is dysregulated in cancer. Myeloid cells are a diverse population of immune cells that are markedly altered in cancer. Comprised of macrophages, dendritic cells (DCs) and myeloid-derived suppressor cells (MDSCs), these significantly contribute to the generation of antitumor immune responses and influence the development and progression of cancer. In order to better understand the role of myeloid cells in the tumor microenvironment, it is necessary to develop and characterize various cancer models. In this review, we will first outline how cancer creates dysfunction in various myeloid cell populations. While evaluating human cells and tissue is ideal for cancer research, there are many limitations to what questions can be investigated. To fill these gaps, there have been several successful murine and canine models that are comparative oncology models to study cancer

---

* Corresponding Author. E-mail: papenfuss.1@osue.edu.

immunology. To understand the potential application of animal models in cancer immunology, this review will also outline some of the benefits and limitations of current animal models and discuss their relevance to studying myeloid cell dysfunction in cancer. Given the importance of myeloid cells in antitumor immune responses and developing efficacious cancer immunotherapies, a better understanding of myeloid cell dysfunction in cancer is necessary, regardless of species.

INTRODUCTION

Cancer is an important disease which contributes to mortality in all species. Over the last century, cancer in both humans and animals has been treated with a combination of surgery, radiation therapy and cytotoxic chemotherapy. While these can often minimize a primary tumor, efficacious therapies for treating residual tumors or metastatic disease are lacking. Additionally, there are often significant side effects of these therapies (e.g., chemotherapy). Thus, the concept of re-educating the body's own immune system to target and kill tumor cells has been an appealing therapeutic approach for over several decades. Such cancer immunotherapy includes antibody treatments, cell-based therapies and cancer vaccines [1-4] that have, unfortunately, met with limited success. This has largely been due to a somewhat limited understanding of how the immune system is dysregulated in cancer patients and ultimately how this impacts the generation and regulation of antitumor immunity [5]. During cancer, the immune system has lost the ability to eliminate cancer cells and has become dysregulated [2, 6]. Although numerous cells and immunosuppressive cascades contribute to immune dysfunction during cancer, this review will focus specifically on myeloid cell dysfunction in cancer and discuss mouse and canine animal models for studying cancer immunology.

MYELOID CELLS IN CANCER

The complex interplay between the tumor and myeloid cell populations contributes to the immune dysregulation limiting both effective antitumor immunity and efficacy of cancer immunotherapies. Myeloid cells are a diverse population of cells that form during hematopoiesis comprised of macrophages, dendritic cells (DCs), granulocytes and their precursors. These myeloid cells play important roles in host defense, inducing adaptive immune responses, promoting inflammation and regulating immune responses. In cancer, the tumor both directly and indirectly affects myeloid cell populations. Such modified myeloid cells then influence innate influence innate and adaptive antitumor immune responses which ultimately results in both local and systemic immunosuppression [7]. Myeloid cell dysfunction is a critical reason that the immune system in cancer fails to function normally and likely contributes to the limited success of cancer immunotherapy. Tumor-associated macrophages (TAMs), abnormal dendritic cells (DCs) and the more recently described myeloid derived suppressor cells (MDSCs) are all myeloid cell dysregulated in cancer.

TAMs

Macrophages have long been known to play an important role in both innate and adaptive immunity. In cancer immunology, there has recently been a strong interest in tumor-associated macrophages (TAMs) and their role in immune dysfunction. TAMs are present within the tumor microenvironment and often derived from blood monocytes or local proliferation. These TAMs exhibit an alternatively activated (i.e., M2-like or anti-inflammatory) phenotype which contributes to immunosuppression and facilitates tumor invasion and metastatic spread [7-12]. TAMs can protect tumors cells from apoptosis, facilitate tumor invasion and metastasis and promote angiogenesis which all contribute to tumor progression [13, 14] [9, 10, 12, 15-18]. TAMs are found in all phases of tumor progression and have reduced antigen-presentation abilities which subsequently limit a T-cell-mediated adaptive immune response. While their role is not fully understood, studies suggest that TAMs may have a significant impact on the success of immunotherapies. It is believed that the immunosuppressive function of TAMs inhibit immunotherapies by enhancing a M2-like tumor promoting response and preventing a M1-like tumorcidial response [19]. TAMs are ineffective antigen-presenting cells and fail to induce anti-tumor immunity. They also may produce immunoregulatory compounds, such as IL-10, TGF-beta and prostaglandin E2 and arginase, can promote the induction of regulatory T cells (T_{regs}) and Th2 cells and also induce apoptosis in effector Th1 cells [20-22]. There are numerous studies investigating the mechanisms involved in generation and pathologic function of various TAM subsets that are beyond the scope of this review. Suffice it to say that there are numerous studies investigating the contribution of TAMs to not only tumor recurrence and metastases, but also how TAMs prevent antitumor immunity and successful cancer immunotherapies.

DCs

Like macrophages, DCs are antigen-presenting cells which bridge innate and adaptive immune responses. DCs are considered unique in their ability to influence naïve helper T cell differentiation and they play an important role in presenting and cross-presenting antigen to both CD4+ and CD8+ T cells [23-30]. DCs are also well equipped to mobilize to lymph nodes and drive antigen-specific immune responses in secondary lymphoid organs such as lymph nodes. In cancer, normal mature functional DCs are decreased with an accumulation of immature DCs and DC precursors (e.g., MDSCs) [7, 27, 28, 31]. By and large, the impaired balance of mature and immature myeloid cells and their relative dysfunction is a hallmark of cancer and progressive tumor growth is associated with a decrease in functionally mature DCs [7, 27, 32, 33]. Such altered DCs may not only have a decreased ability to promote antitumor (i.e., Th1) antigen-specific immune responses, but may promote immune tolerance through the production of immunosuppressive molecules (e.g., TGF-beta, IL-10, indoleamine 2,3 deoxygeanse (IDO)) and induction of various regulatory immune cells [4, 34] [7, 35-43]. Interestingly, although cancer can modify DC function, what is of probably of equal of even greater interest is the target or use DCs to promote effective antitumor immunity. The potent ability of DCs to induce targeted tumor-specific CD4 and CD8 adaptive immune responses is the subject of numerous studies and therapeutic applications [2-4]. Additionally, like most vaccines, cancer vaccines largely target DCs in order to promote an antitumor immune response and DC-targeted vaccines and immunotherapies are the focus of many studies and reviewed elsewhere [44-50]. Creative applications such as DC/tumor fusion cells or targeting of DCs with nanoparticle delivery of antigen or immunostimulants are ongoing areas of investigation. For example, nanovaccines in which nanoparticles are loaded with tumor-specific antigens and then phagocytosed by DCs have been shown to be quite successful at eliciting an anti-tumor response [46]. When these pre-programmed DCs are introduced into cancer patients they elicit a T cell response that is specific to the tumor cells and causes the cells to lyse. Such DC-targeting vaccines have gathered much attention because of their precision in targeting metastatic disease more effectively than either chemotherapy or radiation therapy. However, despite the promise of DC vaccines, there is still much are still many gaps that need to be filled in understanding DC dysfunction and therapeutic applications in cancer. Utilizing and developing new methods and models in the lab to investigate these mechanisms will be crucial to the success of these vaccine treatments and other cancer immunotherapies in the future.

GRANULOCYTES

Granulocytes within the tumor environment largely consist of neutrophils and are considered to be recruited to the tumor environment largely in response to inflammation and/or cellular necrosis present within the tumor. However, some recent studies do suggest that neutrophils are influenced by the tumor environment and may shift from pro-tumoral N-1 phenotype to a pro-tumoral N-2 phenotype [51, 52]. Although TGF-beta is thought to promote this conversion and pro-tumoral N-2s may produce IL-10 and suppres CD8 T cells, the relative impact of neutrophil subtypes of tumor progression and overall immune dysregulation in cancer remains to be explored [52-54].

MDSCs

MDSCs are a recently described cell which has been shown to significantly contribute to the immune dysfunction in cancer. These immunoregulatory cells are important roadblocks to achieving complete anti-tumor immunity and effective cancer immunotherapy. MDSCs represent a myeloid precursor population that, normally, has the potential to differentiate into granulocytes, macrophages or DCs [55, 56]. Thus, regardless of species, these typically co-express myeloid lineage differentiation markers consistent with a myeloid precursor [56]. Specifically, murine MDSCs express the granulocyte lineage marker Gr1 and myeloid CD11b, human MDSCs have the phenotype of CD14-CD11b+HLA-DR- and canine MDSCs with phenotypes consistent with either a murine or human phenotype have been described [57-59] [60, 61]. Numerous MDSC subsets have been defined based on the additional cell surface markers but can broadly divided into MDSCs with more monocyte versus granulocyte phenotype/function which may suppress immune responses through a variety of mechanisms [56, 62].

MDSCs are induced by various factors produced by the tumor environment and/or inflammatory cascades. Factors such as GM-CSF, G-CSF, IL-6, IL-13, S100A8/9 and numerous others promote the expansion and differentiation arrest that leads to the accumulation of MDSCs in a two-step process [56, 62]. Transcription factors such as STAT3 are upregulated in early stages promoting myelopoiesis and MDSC accumulation following by a second activation step required for full regulatory effects of MDSCs to manifest. MDSCs are able to impact numerous other immune cells through a wide array of both direct and indirect mechanisms. MDSCs produce reactive oxygen species and peroxynitrite, deplete local nutrients L-arginine and L-cysteine required for effective lymphocyte function, produce regulatory molecules (e.g., IL-10, TGF-beta and IDO) and express regulatory costimulatory molecules that all serve to modify immune responses [56, 62-65]. MDSCs inhibit the activity of NK and T cells (CD4+ and CD8+), dendritic cells and promote regulatory cells including regulatory T cells (T_{regs}) and regulatory macrophages (e.g., TAMs) [7, 56, 62, 66, 67]. The ultimate result is the inhibition of both innate and adaptive antitumor responses.

There is much interest in diminishing the immunosuppressive effects of MDSCs [68] while enhancing the activity of macrophages and DCs to promote antitumor immunity and improve the efficacy of cancer immunotherapies [68]. However, simple targeting of a single population of myeloid cells is likely to insufficient due to the abundant cross-talk between myeloid cell populations and other cells of the immune system [7]. Although mouse studies have provided much of our current understanding of immune dysregulation in cancer, future studies investigating myeloid and immune cell dysfunction in patient populations will be necessary to develop effective cancer immunotherapies for treating cancer. The remainder of this review will briefly outline some of the key factors to consider in utilizing animal models for studying myeloid cell dysfunction in cancer.

HUMAN CELL LINES AND XENOGRAFT MODELS

In order to study the relationship between myeloid cells and cancer, there have been various models developed to study human cancer. The majority of these models are mouse models and studies have led to a greater understanding of myeloid cell and overall immune response changes that occur in cancer [69-

72]. Ideally, the most relevant means to understand immune alterations and responses to cancer in humans is to evaluate human cancer patients. However, there are obvious limitations to what information can be obtained from such studies. While evaluation of myeloid cell populations in human patients can provide insight and potentially correlate with prognoses, by and large a mechanistic understanding of the cancer-immune cell interactions in a human system has been determined using experimental models. Immortalized human cancer cell lines are commonly used to screen potential anti-cancer drugs but with any cell lines and *in vitro* models, there are potential caveats and the clinical relevance of these cancer cell lines can be called into question [73-78]. Primarily, the responses of an organism cannot be recapitulated in cell culture. The *in vivo* biological relevance and interactions between cells, tissues, organs and tumor is a necessary component to evaluate cancer in its entirety. Animal models are used to investigate these questions and in cancer studies, xenografts of human tissue implanted into immunodeficient mice are largely used [78-80]. Numerous immunodeficient mice are available depending upon the question of interest. Nude, SCID and Rag1 deficient mice are the most commonly used imunodeficient strains available and have varying immunodeficiencies in the T, B and innate (e.g., NK cells, macrophage and complement) immune cell compartments. Nude mice have a disruption in the Foxn1 gene and typically have T cell deficiencies with partial B cell defects. Nude mice have relative intact B cell function and some extrathymic T cell function but are generally less "leaky" compared to SCID mice. SCID mice have genetic disruption of the protein kinase *Prkdc* while Rag1 have a disruption in the recombination activating gene 1. Both of these strains tend to have more severe defects in both the T and B cell compartments. SCID mice are considered "leaky" in that after approximately 12 weeks of age, some functional T and B cells may develop and Rag1 knock-out mice are less well characterized for xenograft studies [81]. Given the loss of both T and B cells, SCID mice are often more commonly used and mouse strains with additional immunodeficiencies (e.g., Scid-beige mices lacking T, B and NK cells or CIEA NOG mice which lack T, B NK cells and have dysfunctional macrophage/ DC/ complement activity, etc.) have been developed [82][83]. Humanized mice have also been explored for applications in cancer research and these immune-altered mice can either have human genes inserted into a mouse genome or reconstituted with human stem cells or lymphocytes [84, 85]. Finally, there are continual advances in creating new and more sophisticated mouse models to answer important questions regarding cancer-immune interactions [69]. Given this large array of possible *in vivo* model system, careful consideration must be given not only to the strain/genetic defect, but also to the anatomic site of implantation given that biological behavior and metastatic potential depends both on the properties of the tumor cells and host factors. For example, innate immunity has been shown to limit tumor growth and prevent metastases in nude mice necessitating the incorporation of NK cell deficiencies for some cancer studies [86]. To date, very few studies have investigated the impact of myeloid cell deficiencies in xenograft models given the non-myeloid cell immunodeficiencies are required to allow engraftment of human tumor cells. Regardless of the model system, the transplantation of human tissue into mice will necessitate some degree of immune modulation and will present inherent limitations in studying cancer-immune interactions [80].

MURINE CANCER MODELS IN IMMUNOCOMPETENT MICE

Obviously, it is difficult to investigate myeloid cell dysfunction, immune alterations or the efficacy of cancer immunotherapy in a model system requiring mice to be immunodeficient. Thus, models using mice with intact immune systems (i.e., immunocompetent) have been used to advance our understanding of cancer immunology. Numerous studies have investigated cancer biology in mice with defined genetic backgrounds and modifications. Such genetically engineered mice (GEMs) have provided an understanding of basic tumor biology and the role of defined genes and pathways in tumor development, progression and metastatic spread [80]. Both spontaneous developing models and induced (either allograft injection of tumor lines or chemical/radiation models can be used for these studies [78, 87, 88]. For example, transgenic mice lacking BRCA1 or BRCA1 genes are commonly used to induce and study breast cancer [88]. Similarly, mice with mutated PTEN tumor suppressor gene have been used to study metastatic disease in prostate cancer and the genetic deletion of such important tumor suppressor genes has provide significant insight into tumor progression [89, 90]. A primary disadvantage to these models is the variable incidence and spontaneous nature of tumor development which limits large cohort studies. Alternatively, induced models allow larger cohort studies to be performed and assessment of therapeutic intervention can more easily be evaluated. However, induction of mouse tumor lines into recipient mice, a tumor of the same genetic background is required and there are relatively few cell lines available on GEMs useful for immunological studies (i.e., mice with a C57Bl/6 and

Balb/c backgrounds). In addition, while some cell lines have potential sublines with varying metastatic/aggressive biological behavior can potentially mimic the variation seen in outbred human populations, not all cell lines have metastatic potential and metastatic models often fail to mimic what is seen in humans [70, 91, 92]. Obviously mice are not men and physiological differences, biomarkers and differences in innate immune responses [72, 93, 94]. However, the obvious advantage of non-xenograft models is that these studies are largely performed in immunocompetent mice or mice with defined genetic mutations which allow for detailed mechanistic studies. Indeed, much of what is currently known regarding immune function and, in particular, myeloid cell changes, have been gleaned from immunocompetent mouse models. Given the intact nature of the host's immune response, mechanistic questions regarding tumor progression, metastasis, drug treatment, or tumor regression can be evaluated from an immunological perspective. Thus, studies in immunocompetent mice studies will likely provide the most significant insight regarding the field of cancer immunology and are important when evaluating the utility of specific cancer immunotherapies. Although these immunocompetent mouse models can be informative regarding myeloid cell dysfunction in cancer, these models still exist within inbred genetic mouse strains so the conclusions in outbred immunologically diverse populations, as seen in humans, is still limited. If the goal is to understand the immune alterations seen in cancer patient populations, additional model systems are necessary.

CANINE AS A MODEL TO STUDY CANCER IMMUNOLOGY

An ideal model will share histopathologic features similar to human tumors, progress through similar stages with physiological effects, involve the same genes and biological pathways and clinically, should be both similar to human responses to a particular therapy and be predictive of therapeutic efficacy in human cancer patients [70, 95]. While rodent (mouse) models have overwhelmingly been the model of choice to understand tumor biology and cancer immunology, there is often a lack of association in the preclinical phase of evaluating cancer therapeutics [96-98]. Additional animal models have been used in a comparative oncology approach to provide information regarding both biological behavior and therapeutic response [99, 100]. Although cancer occurs in all species, dogs are the primary non-rodent species studied by comparative oncologists and studies over the last 30 years have shown the utility of using pet dogs with cancer [95, 96, 100-104]. Dogs have been shown to provide critical information regarding pharmacokinetics and pharmacodynamics, imaging, therapeutic index and efficacy endpoints [102, 103, 105-107]. Studies in pet dogs have also modeled toxicities of drugs in humans and evaluated long-term exposure outcomes of a new drug that conventional strategies couldn't model before first-in human studies were performed [106, 108]. The utility of a comparative oncology approach is recognized and infrastructural support through the Comparative Oncology Trials Consortium at the National Institutes of Health and the Canine comparative oncology & genomics consortium (www.ccogc.net) are now in place [96, 109]. Human tumors have been found to be more similar to dog tumors than mouse tumors [102, 107, 110-112]. Indeed, many canine tumors are reflective counterparts to those seen in humans and include; non-Hodgkin lymphoma, osteosarcoma, glioblastoma multiforme, melanoma, prostate carcinoma, lung carcinoma, head & neck carcinoma, mammary carcinoma and soft-tissue sarcoma and these tumor types are very similar to human cancers in regards to histology, tumor genetics, biological behavior and response to conventional therapies [101-103, 107, 112-115]. Additionally, many of these effectively model metastatic disease which mouse models often fail to do [116]. While significant advances have been made in our understanding of the genetics, biological behavior and modeling potential of canine tumors, less is known about canine immune responses to cancer.

Unlike inbred mouse models, dogs are outbred like humans and largely exposed to similar environmental conditions as humans. Additionally, dogs and humans are similar genetically and physiologically, have relatively long lifespans compared to mice, have similar cancer progression and metastatic profiles to humans and canine cancers tend to respond to radiation and chemotherapies similar to what is seen in humans [100, 102, 103, 107]. With regard to evaluating new drug candidate, pet dogs with cancer are increasingly being used in both preclinical and clinical components of drug development. This is due not only to the similarities cited above, but also due to facts that pet dogs are not subject to the constraints (e.g., FDA approval, etc.) of human clinical trials and that traditional human cancer therapies are often investigational by nature when used in veterinary medicine. Increasingly, preclinical testing for new cancer drugs are being investigated in the pet dog population and answers such as toxicity, dose, regimen, pharmacokinetics, pharmacodynamics and activity can be answered before being used in humans. For example, tyrosine kinase inhibitors studies in dogs demonstrated antitumor activity and predicted toxicities and these were performed before testing in human patients [106, 108]. Additionally, the availability of canine cancer cell lines can facilitate cross-species comparisons to human cell lines regarding the evaluation of new and novel drug or immunotherapy candidates [117-121]. It is somewhat ironic that canine cancer patients, that had previously had to wait for approval and availability of human anti-cancer drugs, are now being the first to evaluate potentially life-saving therapies.

Although in its infancy, cancer immunotherapy is being explored in canine cancer patients [122-124]. Some specific examples include xenogeic vaccines for canine melanoma, RNA-loaded CD40-activated B cells stimulate antigen-specific T-cell responses in dogs with spontaneous lymphoma, the use of the immunomodulatory MTP-PE in osteosarcoma and glioblastoma treatment adenoviral vectors have been investigated [106, 113, 122, 125-127]. The xenogeneic DNA vaccine, OnCept^{TM} is now a USDA-approved treatment for melanoma in dogs and xenogeneic DNA vaccination is being evaluated in human melanoma patients [122, 128-130]. Although immunotherapies are being investigated in veterinary medicine, there still remains much to be learned regarding canine immune responses. In many cases, immune read-outs are limited in scope, perhaps due to limited availability of canine-specific immune reagents. When immune responses are evaluated, canine cancer patients model the findings in human cancer patients and demonstrate that cancer in dogs modifies immune responses. For example, levels of immunoregulatory FoxP3+ T_{regs} are increased in cancer and correlate with metastatic burden [131-136] and therapies can modify these and overall immune responses. Most recently, we and others have shown that myeloid cell responses can be altered in cancer although few additional studies have specifically investigated myeloid cell dysfunction in cancer [60, 61, 119, 137-139]. If dogs are to useful in developing cancer immunotherapies and understanding immune responses to cancer, a more thorough understanding of canine immune responses, particularly myeloid cell dysfunction is necessary.

CONCLUSION

In summary, this review provides basic background of information regarding myeloid cell dysfunction in cancer and how such altered myeloid cells may impact antitumor immune responses and the efficacy of cancer immunotherapy. Given that direct evaluation of immune in response to cancer has limitations, this review discusses some of the considerations regarding animal model use to study cancer immunology. While inbred mouse models have provided much of our basic understanding of how cancer modified myeloid cell and overall immune system dysfunction, there are significant limitations to the various mouse model systems currently available. Pet dogs with cancer provide a promising and potentially more relevant model for evaluating both cancer biology and cancer therapeutics, our understanding of canine immune responses, particularly myeloid cells, is severely lacking. Myeloid cells are critical for the development of antitumor immune responses and are central to many cancer immunotherapies. Given their importance, a better understanding of myeloid cell dysfunction in cancer, regardless of species, is critical.

REFERENCES

[1]   Vesely, M. D. and Schreiber, R. D. (2013) Cancer immunoediting: antigens, mechanisms, and implications to cancer immunotherapy. *Annals of the New York Academy of Sciences* 1284, 1-5.

[2]   Mellman, I., Coukos, G., Dranoff, G. (2011) Cancer immunotherapy comes of age. *Nature* 480, 480-9.

[3]   Topalian, S. L., Weiner, G. J., Pardoll, D. M. (2011) Cancer immunotherapy comes of age. *Journal of clinical oncology : official journal of the American Society of Clinical Oncology* 29, 4828-36.

[4]   Schuler, G., Schuler-Thurner, B., Steinman, R. M. (2003) The use of dendritic cells in cancer immunotherapy. *Current opinion in immunology* 15, 138-47.

[5]   Kawakami, Y., Yaguchi, T., Sumimoto, H., Kudo-Saito, C., Tsukamoto, N., Iwata-Kajihara, T., Nakamura, S., Nishio, H., Satomi, R., Kobayashi, A., Tanaka, M., Park, J. H., Kamijuku, H., Tsujikawa, T., Kawamura, N. (2013) Cancer-induced immunosuppressive cascades and their reversal by molecular-targeted therapy. *Annals of the New York Academy of Sciences* 1284, 80-6.

[6]   Vesely, M. D., Kershaw, M. H., Schreiber, R. D., Smyth, M. J. (2011) Natural innate and adaptive immunity to cancer. *Annual review of immunology* 29, 235-71.

[7]   Gabrilovich, D. I., Ostrand-Rosenberg, S., Bronte, V. (2012) Coordinated regulation of myeloid cells by tumours. *Nature reviews* 12, 253-68.

[8]   De Palma, M. and Lewis, C. E. (2013) Macrophage regulation of tumor responses to anticancer therapies. *Cancer cell* 23, 277-86.

[9]   Mantovani, A. and Sica, A. (2010) Macrophages, innate immunity and cancer: balance, tolerance, and diversity. *Current opinion in immunology* 22, 231-7.

[10]   Mantovani, A., Sica, A., Allavena, P., Garlanda, C., Locati, M. (2009) Tumor-associated macrophages and the related myeloid-derived suppressor cells as a paradigm of the diversity of macrophage activation. *Human immunology* 70, 325-330.

[11]   Allavena, P., Sica, A., Garlanda, C., Mantovani, A. (2008) The Yin-Yang of tumor-associated macrophages in neoplastic progression and immune surveillance. *Immunological reviews* 222, 155-61.

[12]   Gordon, S. and Martinez, F. O. (2010) Alternative activation of macrophages: mechanism and functions. *Immunity* 32, 593-604.

[13]   Zheng, Y., Cai, Z., Wang, S., Zhang, X., Qian, J., Hong, S., Li, H., Wang, M., Yang, J., Yi, Q. (2009) Macrophages are an abundant component of myeloma microenvironment and protect myeloma cells from chemotherapy drug-induced apoptosis. *Blood* 114, 3625-8.

[14]   Qian, B. Z., Deng, Y., Im, J. H., Muschel, R. J., Zou, Y. Y., Li, J. F., Lang, R. A., Pollard, J. W. (2009) A Distinct Macrophage Population Mediates Metastatic Breast Cancer Cell Extravasation, Establishment and Growth. *Plos One* 4.

[15]   Lin, E. Y., Li, J. F., Gnatovskiy, L., Deng, Y., Zhu, L., Grzesik, D. A., Qian, H., Xue, X. N., Pollard, J. W. (2006) Macrophages regulate the angiogenic switch in a mouse model of breast cancer. *Cancer research* 66, 11238-46.

[16]   Qian, B. Z. and Pollard, J. W. (2010) Macrophage diversity enhances tumor progression and metastasis. *Cell* 141, 39-51.

[17]   Biswas, S. K., Chittezhath, M., Shalova, I. N., Lim, J. Y. (2012) Macrophage polarization and plasticity in health and disease. *Immunol. Res.* 53, 11-24.

[18]   Biswas, S. K. and Mantovani, A. (2010) Macrophage plasticity and interaction with lymphocyte subsets: cancer as a paradigm. *Nat. Immunol.* 11, 889-896.

[19]   Tang, X., Mo, C., Wang, Y., Wei, D., Xiao, H. (2013) Anti-tumour strategies aiming to target tumour-associated macrophages. *Immunology* 138, 93-104.

[20] Kuang, D. M., Zhao, Q. Y., Peng, C., Xu, J., Zhang, J. P., Wu, C. Y., Zheng, L. M. (2009) Activated monocytes in peritumoral stroma of hepatocellular carcinoma foster immune privilege and disease progression through PD-L1. *J. Exp. Med.* 206, 1327-1337.

[21] Rodriguez, P. C., Quiceno, D. G., Zabaleta, J., Ortiz, B., Zea, A. H., Piazuelo, M. B., Delgado, A., Correa, P., Brayer, J., Sotomayor, E. M., Antonia, S., Ochoa, J. B., Ochoa, A. C. (2004) Arginase I production in the tumor microenvironment by mature myeloid cells inhibits T-cell receptor expression and antigen-specific T-cell responses. *Cancer research* 64, 5839-5849.

[22] Torroella-Kouri, M., Silvera, R., Rodriguez, D., Caso, R., Shatry, A., Opiela, S., Ilkovitch, D., Schwendener, R. A., Iragavarapu-Charyulu, V., Cardentey, Y., Strbo, N., Lopez, D. M. (2009) Identification of a subpopulation of macrophages in mammary tumor-bearing mice that are neither M1 nor M2 and are less differentiated. *Cancer research* 69, 4800-9.

[23] Mellman, I. and Steinman, R. M. (2001) Dendritic cells: specialized and regulated antigen processing machines. *Cell* 106, 255-8.

[24] Banchereau, J. and Steinman, R. M. (1998) Dendritic cells and the control of immunity. *Nature* 392, 245-52.

[25] Steinman, R. M. (2012) Decisions about dendritic cells: past, present, and future. *Annual review of immunology* 30, 1-22.

[26] Melief, C. J. (2003) Mini-review: Regulation of cytotoxic T lymphocyte responses by dendritic cells: peaceful coexistence of cross-priming and direct priming? *European journal of immunology* 33, 2645-54.

[27] Gabrilovich, D. (2004) Mechanisms and functional significance of tumour-induced dendritic-cell defects. *Nature reviews* 4, 941-52.

[28] van der Bruggen, P. and Van den Eynde, B. J. (2006) Processing and presentation of tumor antigens and vaccination strategies. *Current opinion in immunology* 18, 98-104.

[29] Lin, M. L., Zhan, Y. F., Villadangos, J. A., Lew, A. M. (2008) The cell biology of cross-presentation and the role of dendritic cell subsets. *Immunol. Cell Biol.* 86, 353-362.

[30] Steinman, R. M., Pack, M., Inaba, K. (1997) Dendritic cells in the T-cell areas of lymphoid organs. *Immunological reviews* 156, 25-37.

[31] Melief, C. J. (2008) Cancer immunotherapy by dendritic cells. *Immunity* 29, 372-83.

[32] Kusmartsev, S. and Gabrilovich, D. I. (2002) Immature myeloid cells and cancer-associated immune suppression. *Cancer Immunol. Immun.* 51, 293-298.

[33] Yurochko, A. D., Pyle, R. H., Elgert, K. D. (1989) Changes in macrophage populations: phenotypic differences between normal and tumor-bearing host macrophages. *Immunobiology* 178, 416-35.

[34] Herber, D. L., Cao, W., Nefedova, Y., Novitskiy, S. V., Nagaraj, S., Tyurin, V. A., Corzo, A., Cho, H. I., Celis, E., Lennox, B., Knight, S. C., Padhya, T., McCaffrey, T. V., McCaffrey, J. C., Antonia, S., Fishman, M., Ferris, R. L., Kagan, V. E., Gabrilovich, D. I. (2010) Lipid accumulation and dendritic cell dysfunction in cancer. *Nature medicine* 16, 880-6.

[35] Novitskiy, S. V., Ryzhov, S., Zaynagetdinov, R., Goldstein, A. E., Huang, Y., Tikhomirov, O. Y., Blackburn, M. R., Biaggioni, I., Carbone, D. P., Feoktistov, I., Dikov, M. M. (2008) Adenosine receptors in regulation of dendritic cell differentiation and function. *Blood* 112, 1822-31.

[36] Dumitriu, I. E., Dunbar, D. R., Howie, S. E., Sethi, T., Gregory, C. D. (2009) Human dendritic cells produce TGF-beta 1 under the influence of lung carcinoma cells and prime the differentiation of CD4+CD25+Foxp3+ regulatory T cells. *Journal of Immunology* 182, 2795-807.

[37] Liu, Q. Y., Zhang, C. X., Sun, A. N., Zheng, Y. Y., Wang, L., Cao, X. T. (2009) Tumor-Educated CD11b(high)Ia(low) Regulatory Dendritic Cells Suppress T Cell Response through Arginase I. *Journal of Immunology* 182, 6207-6216.

[38] Watkins, S. K., Zhu, Z., Riboldi, E., Shafer-Weaver, K. A., Stagliano, K. E., Sklavos, M. M., Ambs, S., Yagita, H., Hurwitz, A. A. (2011) FOXO3 programs tumor-associated DCs to become tolerogenic in human and murine prostate cancer. *The Journal of clinical investigation* 121, 1361-72.

[39] Munn, D. H., Sharma, M. D., Hou, D., Baban, B., Lee, J. R., Antonia, S. J., Messina, J. L., Chandler, P., Koni, P. A., Mellor, A. L. (2004) Expression of indoleamine 2,3-dioxygenase by plasmacytoid dendritic cells in tumor-draining lymph nodes. *The Journal of clinical investigation* 114, 280-90.

[40] Lee, J. R., Dalton, R. R., Messina, J. L., Sharma, M. D., Smith, D. M.,
Burgess, R. E., Mazzella, F., Antonia, S. J., Mellor, A. L., Munn, D. H.
(2003) Pattern of recruitment of immunoregulatory antigen-presenting
cells in malignant melanoma. *Laboratory investigation; a journal of
technical methods and pathology* 83, 1457-66.

[41] Shurin, G. V., Ma, Y., Shurin, M. R. (2013) Immunosuppressive
Mechanisms of Regulatory Dendritic Cells in Cancer. *Cancer
microenvironment: official journal of the International Cancer
Microenvironment Society.*

[42] Ma, Y., Shurin, G. V., Peiyuan, Z., Shurin, M. R. (2013) Dendritic cells
in the cancer microenvironment. *Journal of Cancer* 4, 36-44.

[43] Steinman, R. M., Hawiger, D., Nussenzweig, M. C. (2003) Tolerogenic
dendritic cells. *Annual review of immunology* 21, 685-711.

[44] Palucka, K. and Banchereau, J. (2012) Cancer immunotherapy via
dendritic cells. Nature reviews. *Cancer* 12, 265-77.

[45] Avigan, D., Rosenblatt, J., Kufe, D. (2012) Dendritic/tumor fusion cells
as cancer vaccines. *Seminars in oncology* 39, 287-95.

[46] Paulis, L. E., Mandal, S., Kreutz, M., Figdor, C. G. (2013) Dendritic
cell-based nanovaccines for cancer immunotherapy. *Current opinion in
immunology* 25, 389-95.

[47] Cruz, L. J., Tacken, P. J., Rueda, F., Domingo, J. C., Albericio, F.,
Figdor, C. G. (2012) Targeting nanoparticles to dendritic cells for
immunotherapy. *Methods in enzymology* 509, 143-63.

[48] Goutagny, N., Estornes, Y., Hasan, U., Lebecque, S., Caux, C. (2012)
Targeting pattern recognition receptors in cancer immunotherapy.
Targeted oncology 7, 29-54.

[49] Dubensky, T. W., Jr. and Reed, S. G. (2010) Adjuvants for cancer
vaccines. *Seminars in immunology* 22, 155-61.

[50] Kaczanowska, S., Joseph, A. M., Davila, E. (2013) TLR agonists: our
best frenemy in cancer immunotherapy. *Journal of leukocyte biology* 93,
847-63.

[51] Granot, Z., Henke, E., Comen, E. A., King, T. A., Norton, L., Benezra,
R. (2011) Tumor Entrained Neutrophils Inhibit Seeding in the
Premetastatic Lung. *Cancer cell* 20, 300-314.

[52] Fridlender, Z. G., Sun, J., Kim, S., Kapoor, V., Cheng, G. J., Ling, L. N.,
Worthen, G. S., Albelda, S. M. (2009) Polarization of Tumor-Associated

Neutrophil Phenotype by TGF-beta: "N1" versus "N2" TAN. *Cancer cell* 16, 183-194.

[53] Davey, M. S., Tamassia, N., Rossato, M., Bazzoni, F., Calzetti, F., Bruderek, K., Sironi, M., Zimmer, L., Bottazzi, B., Mantovani, A., Brandau, S., Moser, B., Eberl, M., Cassatella, M. A. (2011) Failure to detect production of IL-10 by activated human neutrophils. *Nat. Immunol.* 12, 1017-8; author reply 1018-20.

[54] De Santo, C., Arscott, R., Booth, S., Karydis, I., Jones, M., Asher, R., Salio, M., Middleton, M., Cerundolo, V. (2010) Invariant NKT cells modulate the suppressive activity of IL-10-secreting neutrophils differentiated with serum amyloid A. *Nat. Immunol.* 11, 1039-46.

[55] Youn, J. I. and Gabrilovich, D. I. (2010) The biology of myeloid-derived suppressor cells: the blessing and the curse of morphological and functional heterogeneity. *European journal of immunology* 40, 2969-75.

[56] Gabrilovich, D. I. and Nagaraj, S. (2009) Myeloid-derived suppressor cells as regulators of the immune system. *Nature reviews* 9, 162-74.

[57] Kusmartsev, S., Nefedova, Y., Yoder, D., Gabrilovich, D. I. (2004) Antigen-specific inhibition of CD8+ T cell response by immature myeloid cells in cancer is mediated by reactive oxygen species. *Journal of Immunology* 172, 989-99.

[58] Ochoa, A. C., Zea, A. H., Hernandez, C., Rodriguez, P. C. (2007) Arginase, prostaglandins, and myeloid-derived suppressor cells in renal cell carcinoma. *Clin. Cancer Res.* 13, 721s-726s.

[59] Almand, B., Clark, J. I., Nikitina, E., van Beynen, J., English, N. R., Knight, S. C., Carbone, D. P., Gabrilovich, D. I. (2001) Increased production of immature myeloid cells in cancer patients: a mechanism of immunosuppression in cancer. *Journal of Immunology* 166, 678-89.

[60] Sherger, M., Kisseberth, W., London, C., Olivo-Marston, S., Papenfuss, T. L. (2012) Identification of myeloid derived suppressor cells in the peripheral blood of tumor bearing dogs. *BMC veterinary research* 8, 209.

[61] Goulart, M. R., Pluhar, G. E., Ohlfest, J. R. (2012) Identification of myeloid derived suppressor cells in dogs with naturally occurring cancer. *Plos One* 7, e33274.

[62] Ostrand-Rosenberg, S. and Sinha, P. (2009) Myeloid-derived suppressor cells: linking inflammation and cancer. *J. Immunol.* 182, 4499-506.

[63] Liu, Y., Zeng, B., Zhang, Z. H., Zhang, Y., Yang, R. C. (2008) B7-H1 on myeloid-derived suppressor cells in immune suppression by a mouse model of ovarian cancer. *Clin. Immunol.* 129, 471-481.

[64] Yu, J. P., Du, W. J., Yan, F., Wang, Y., Li, H., Cao, S., Yu, W. W., Shen, C., Liu, J. T., Ren, X. B. (2013) Myeloid-Derived Suppressor Cells Suppress Antitumor Immune Responses through IDO Expression and Correlate with Lymph Node Metastasis in Patients with Breast Cancer. *Journal of Immunology* 190, 3783-3797.

[65] Ostrand-Rosenberg, S. (2010) Myeloid-derived suppressor cells: more mechanisms for inhibiting antitumor immunity. *Cancer immunology, immunotherapy*: CII 59, 1593-600.

[66] Li, H. Q., Han, Y. M., Guo, Q. L., Zhang, M. G., Cao, X. T. (2009) Cancer-Expanded Myeloid-Derived Suppressor Cells Induce Anergy of NK Cells through Membrane-Bound TGF-beta 1. *Journal of Immunology* 182, 240-249.

[67] Srivastava, M. K., Sinha, P., Clements, V. K., Rodriguez, P., Ostrand-Rosenberg, S. (2010) Myeloid-derived suppressor cells inhibit T-cell activation by depleting cystine and cysteine. *Cancer research* 70, 68-77.

[68] Ugel, S., Delpozzo, F., Desantis, G., Papalini, F., Simonato, F., Sonda, N., Zilio, S., Bronte, V. (2009) Therapeutic targeting of myeloid-derived suppressor cells. *Current opinion in pharmacology* 9, 470-81.

[69] Cheon, D. J. and Orsulic, S. (2011) Mouse Models of Cancer. *Annu. Rev. Pathol-Mech.* 6, 95-119.

[70] Cespedes, M. V., Casanova, I., Parreno, M., Mangues, R. (2006) Mouse models in oncogenesis and cancer therapy. Clinical & translational oncology: official publication of the Federation of Spanish Oncology Societies and of the National Cancer Institute of Mexico 8, 318-29.

[71] Frese, K. K. and Tuveson, D. A. (2007) Maximizing mouse cancer models. *Nature Reviews Cancer* 7, 645-658.

[72] Abate-Shen, C. (2006) A new generation of mouse models of cancer for translational research. *Clin. Cancer Res.* 12, 5274-6.

[73] Weinstein, J. N. (2012) DRUG DISCOVERY Cell lines battle cancer. *Nature* 483, 544-545.

[74] Sharpless, N. E. and DePinho, R. A. (2006) Model organisms - The mighty mouse: genetically engineered mouse models in cancer drug development. *Nat. Rev. Drug Discov.* 5, 741-754.

[75] Borrell, B. (2010) How accurate are cancer cell lines? *Nature* 463, 858-858.

[76] Gillet, J. P., Varma, S., Gottesman, M. M. (2013) The Clinical Relevance of Cancer Cell Lines. *Jnci-J. Natl. Cancer I.* 105, 452-458.

[77] Wagner, K. U. (2004) Models of breast cancer: quo vadis, animal modeling? *Breast cancer research: BCR* 6, 31-8.

[78] Talmadge, J. E., Singh, R. K., Fidler, I. J., Raz, A. (2007) Murine models to evaluate novel and conventional therapeutic strategies for cancer. *Am. J. Pathol.* 170, 793-804.

[79] Mattern, J., Bak, M., Hahn, E. W., Volm, M. (1988) Human-Tumor Xenografts as Model for Drug-Testing. *Cancer Metast. Rev.* 7, 263-284.

[80] Richmond, A. and Su, Y. J. (2008) Mouse xenograft models vs GEM models for human cancer therapeutics. *Dis. Model Mech.* 1, 78-82.

[81] Communication, J. (2000) Immunodeficient Model Selection: Choosing a nude, scid or Rag1 strain. In Scientific Publication for Users of JAX Mice *JAX Mice*.

[82] Xie, X., Brunner, N., Jensen, G., Albrectsen, J., Gotthardsen, B., Rygaard, J. (1992) Comparative-Studies between Nude and Scid Mice on the Growth and Metastatic Behavior of Xenografted Human Tumors. *Clin. Exp. Metastas* 10, 201-210.

[83] Taconic (2013) CIEA NOG Mouse. In *Products and Services: Taconic Transgenic Exchange*.

[84] Legrand, N., Weijer, K., Spits, H. (2006) Experimental models to study development and function of the human immune system in vivo. *Journal of Immunology* 176, 2053-2058.

[85] Brehm, M. A., Shultz, L. D., Greiner, D. L. (2010) Humanized mouse models to study human diseases. *Curr. Opin. Endocrinol.* 17, 120-125.

[86] Habu, S., Fukui, H., Shimamura, K., Kasai, M., Nagai, Y., Okumura, K., Tamaoki, N. (1981) Invivo Effects of Anti-Asialo Gm1 .1. Reduction of Nk Activity and Enhancement of Transplanted Tumor-Growth in Nude-Mice. *Journal of Immunology* 127, 34-38.

[87] Imaoka, T., Nishimura, M., Iizuka, D., Daino, K., Takabatake, T., Okamoto, M., Kakinuma, S., Shimada, Y. (2009) Radiation-induced mammary carcinogenesis in rodent models: what's different from chemical carcinogenesis? *Journal of radiation research* 50, 281-93.

[88] Borowsky, A. D. (2011) Choosing a mouse model: experimental biology in context--the utility and limitations of mouse models of breast cancer. *Cold Springs Harbor Perspectives in Biology* 3, a009670.

[89] Wang, S. Y., Gao, J., Lei, Q. Y., Rozengurt, N., Pritchard, C., Jiao, J., Thomas, G. V., Li, G., Roy-Burman, P., Nelson, P. S., Liu, X., Wu, H. (2003) Prostate-specific deletion of the murine Pten tumor suppressor gene leads to metastatic prostate cancer. *Cancer cell* 4, 209-221.

[90] Carver, B. S. and Pandolfi, P. P. (2006) Mouse Modeling in oncologic preclinical and translational research. *Clin. Cancer Res.* 12, 5305-5311.

[91] Aslakson, C. J. and Miller, F. R. (1992) Selective events in the metastatic process defined by analysis of the sequential dissemination of subpopulations of a mouse mammary tumor. *Cancer research* 52, 1399-405.

[92] Ostrand-Rosenberg, S. (2004) Animal models of tumor immunity, immunotherapy and cancer vaccines. *Current opinion in immunology* 16, 143-50.

[93] Seok, J., Warren, H. S., Cuenca, A. G., Mindrinos, M. N., Baker, H. V., Xu, W., Richards, D. R., McDonald-Smith, G. P., Gao, H., Hennessy, L., Finnerty, C. C., Lopez, C. M., Honari, S., Moore, E. E., Minei, J. P., Cuschieri, J., Bankey, P. E., Johnson, J. L., Sperry, J., Nathens, A. B., Billiar, T. R., West, M. A., Jeschke, M. G., Klein, M. B., Gamelli, R. L., Gibran, N. S., Brownstein, B. H., Miller-Graziano, C., Calvano, S. E., Mason, P. H., Cobb, J. P., Rahme, L. G., Lowry, S. F., Maier, R. V., Moldawer, L. L., Herndon, D. N., Davis, R. W., Xiao, W., Tompkins, R. G. (2013) Genomic responses in mouse models poorly mimic human inflammatory diseases. *Proceedings of the National Academy of Sciences of the United States of America* 110, 3507-12.

[94] de Jong, M. and Maina, T. (2010) Of Mice and Humans: Are They the Same?-Implications in Cancer Translational Research. *J. Nucl. Med.* 51, 501-504.

[95] Porrello, A., Cardelli, P., Spugnini, E. P. (2006) Oncology of companion animals as a model for humans. an overview of tumor histotypes. *Journal of experimental & clinical cancer research*: CR 25, 97-105.

[96] Gordon, I., Paoloni, M., Mazcko, C., Khanna, C. (2009) The Comparative Oncology Trials Consortium: Using Spontaneously Occurring Cancers in Dogs to Inform the Cancer Drug Development Pathway. *Plos Med.* 6.

[97] Johnson, J. I., Decker, S., Zaharevitz, D., Rubinstein, L. V., Venditti, J., Schepartz, S., Kalyandrug, S., Christian, M., Arbuck, S., Hollingshead, M., Sausville, E. A. (2001) Relationships between drug activity in NCI preclinical in vitro and in vivo models and early clinical trials. *Brit. J. Cancer* 84, 1424-1431.

[98] Peterson, J. K. and Houghton, P. J. (2004) Integrating pharmacology and in vivo cancer models in preclinical and clinical drug development. *Eur. J. Cancer* 40, 837-44.

[99] Hansen, K. and Khanna, C. (2004) Spontaneous and genetically engineered animal models: use in preclinical cancer drug development. *Eur. J. Cancer* 40, 858-880.

[100] Gordon, I. K. and Khanna, C. (2010) Modeling Opportunities in Comparative Oncology for Drug Development. *Ilar. J.* 51, 214-220.

[101] Pinho, S. S., Carvalho, S., Cabral, J., Reis, C. A., Gartner, F. (2012) Canine tumors: a spontaneous animal model of human carcinogenesis. *Transl. Res.* 159, 165-172.

[102] Paoloni, M. and Khanna, C. (2008) Science and society - Translation of new cancer treatments from pet dogs to humans. *Nature Reviews Cancer* 8, 147-156.

[103] Paoloni, M. C. and Khanna, C. (2007) Comparative oncology today. *Vet. Clin. N. Am-Small* 37, 1023-+.

[104] Vail, D. M. and MacEwen, E. G. (2000) Spontaneously occurring tumors of companion animals as models for human cancer. *Cancer Invest.* 18, 781-792.

[105] Khanna, C., London, C., Vail, D., Mazcko, C., Hirschfeld, S. (2009) Guiding the optimal translation of new cancer treatments from canine to human cancer patients. *Clin. Cancer Res.* 15, 5671-7.

[106] Liao, A. T., McMahon, M., London, C. (2005) Characterization, expression and function of c-Met in canine spontaneous cancers. *Veterinary and comparative oncology* 3, 61-72.

[107] Paoloni, M. and Khanna, C. (2008) Translation of new cancer treatments from pet dogs to humans. Nature reviews. *Cancer* 8, 147-56.

[108] London, C. A., Hannah, A. L., Zadovoskaya, R., Chien, M. B., Kollias-Baker, C., Rosenberg, M., Downing, S., Post, G., Boucher, J., Shenoy, N., Mendel, D. B., McMahon, G., Cherrington, J. M. (2003) Phase I dose-escalating study of SU11654, a small molecule receptor tyrosine

kinase inhibitor, in dogs with spontaneous malignancies. *Clin. Cancer Res.* 9, 2755-68.

[109] Paoloni, M., Lana, S., Thamm, D., Mazcko, C., Withrow, S. (2010) The creation of the Comparative Oncology Trials Consortium Pharmacodynamic Core: Infrastructure for a virtual laboratory. *Vet. J.* 185, 88-89.

[110] Breen, M. (2009) Update on Genomics in Veterinary Oncology. *Top Companion Anim.* M 24, 113-121.

[111] Breen, M. and Modiano, J. F. (2008) Evolutionarily conserved cytogenetic changes in hematological malignancies of dogs and humans - man and his best friend share more than companionship. *Chromosome Res.* 16, 145-154.

[112] Withrow, S. J. and Wilkins, R. M. (2010) Cross talk from pets to people: translational osteosarcoma treatments. *Ilar. J.* 51, 208-13.

[113] Candolfi, M., Curtin, J. F., Nichols, W. S., Muhammad, A. G., King, G. D., Pluhar, G. E., McNiel, E. A., Ohlfest, J. R., Freese, A. B., Moore, P. F., Lerner, J., Lowenstein, P. R., Castro, M. G. (2007) Intracranial glioblastoma models in preclinical neuro-oncology: neuropathological characterization and tumor progression. *Journal of neuro-oncology* 85, 133-48.

[114] Lipsitz, D., Higgins, R. J., Kortz, G. D., Dickinson, P. J., Bollen, A. W., Naydan, D. K., LeCouteur, R. A. (2003) Glioblastoma multiforme: clinical findings, magnetic resonance imaging, and pathology in five dogs. *Veterinary pathology* 40, 659-69.

[115] Mueller, F., Fuchs, B., Kaser-Hotz, B. (2007) Comparative biology of human and canine osteosarcoma. *Anticancer research* 27, 155-64.

[116] Khanna, C. and Hunter, K. (2005) Modeling metastasis in vivo. *Carcinogenesis* 26, 513-523.

[117] Liao, A. T., McCleese, J., Kamerling, S., Christensen, J., London, C. A. (2007) A novel small molecule Met inhibitor, PF2362376, exhibits biological activity against osteosarcoma. *Veterinary and comparative oncology* 5, 177-96.

[118] Hosoya, K., Murahari, S., Laio, A., London, C. A., Couto, C. G., Kisseberth, W. C. (2008) Biological activity of dihydroartemisinin in canine osteosarcoma cell lines. *American journal of veterinary research* 69, 519-26.

[119] Wasserman, J., Diese, L., VanGundy, Z., London, C., Carson, W. E., Papenfuss, T. L. (2012) Suppression of canine myeloid cells by soluble factors from cultured canine tumor cells. *Veterinary immunology and immunopathology* 145, 420-30.

[120] McMahon, M. B., Bear, M. D., Kulp, S. K., Pennell, M. L., London, C. A. (2010) Biological activity of gemcitabine against canine osteosarcoma cell lines in vitro. *American journal of veterinary research* 71, 799-808.

[121] Couto, J. I., Bear, M. D., Lin, J., Pennel, M., Kulp, S. K., Kisseberth, W. C., London, C. A. (2012) Biologic activity of the novel small molecule STAT3 inhibitor LLL12 against canine osteosarcoma cell lines. *BMC Vet. Res.* 8, 244.

[122] Bergman, P. J. (2010) Cancer Immunotherapy. *Vet. Clin. N. Am-Small* 40, 507-+.

[123] Chung, D. S., Kim, C. H., Hong, Y. K. (2012) Animal models for vaccine therapy. *Advances in experimental medicine and biology* 746, 143-50.

[124] London, C. A. (2007) The role of small molecule inhibitors for veterinary patients. *Vet. Clin. N. Am-Small* 37, 1121-+.

[125] Mason, N. J., Coughlin, C. M., Overley, B., Cohen, J. N., Mitchell, E. L., Colligon, T. A., Clifford, C. A., Zurbriggen, A., Sorenmo, K. U., Vonderheide, R. H. (2008) RNA-loaded CD40-activated B cells stimulate antigen-specific T-cell responses in dogs with spontaneous lymphoma. *Gene therapy* 15, 955-65.

[126] Kozicki, A. R., Robat, C., Chun, R., Kurzman, I. D. (2013) Adjuvant therapy with carboplatin and pamidronate for canine appendicular osteosarcoma. *Veterinary and comparative oncology*.

[127] Kurzman, I. D., MacEwen, E. G., Rosenthal, R. C., Fox, L. E., Keller, E. T., Helfand, S. C., Vail, D. M., Dubielzig, R. R., Madewell, B. R., Rodriguez, C. O., Jr., et al. (1995) Adjuvant therapy for osteosarcoma in dogs: results of randomized clinical trials using combined liposome-encapsulated muramyl tripeptide and cisplatin. *Clin. Cancer Res.* 1, 1595-601.

[128] Bergman, P. J., Camps-Palau, M. A., McKnight, J. A., Leibman, N. F., Craft, D. M., Leung, C., Liao, J., Riviere, I., Sadelain, M., Hohenhaus, A. E., Gregor, P., Houghton, A. N., Perales, M. A., Wolchok, J. D. (2006) Development of a xenogeneic DNA vaccine program for canine

malignant melanoma at the Animal Medical Center. *Vaccine* 24, 4582-4585.

[129] Bergman, P. J. and Wolchok, J. D. (2008) Of mice and men (and dogs): development of a xenogeneic DNA vaccine for canine oral malignant melanoma. *Cancer Therapy* 6, 817-826.

[130] Ginsberg, B. A., Gallardo, H. F., Rasalan, T. S., Adamow, M., Mu, Z., Tandon, S., Bewkes, B. B., Roman, R. A., Chapman, P. B., Schwartz, G. K., Carvajal, R. D., Panageas, K. S., Terzulli, S. L., Houghton, A. N., Yuan, J. D., Wolchok, J. D. (2010) Immunologic response to xenogeneic gp100 DNA in melanoma patients: comparison of particle-mediated epidermal delivery with intramuscular injection. *Clin. Cancer Res.* 16, 4057-65.

[131] Horiuchi, Y., Tominaga, M., Ichikawa, M., Yamashita, M., Jikumaru, Y., Nariai, Y., Nakajima, Y., Kuwabara, M., Yukawa, M. (2009) Increase of regulatory T cells in the peripheral blood of dogs with metastatic tumors. *Microbiology and immunology* 53, 468-74.

[132] Horiuchi, Y., Tominaga, M., Ichikawa, M., Yamashita, M., Okano, K., Jikumaru, Y., Nariai, Y., Nakajima, Y., Kuwabara, M., Yukawa, M. (2010) Relationship between regulatory and type 1 T cells in dogs with oral malignant melanoma. *Microbiology and immunology* 54, 152-9.

[133] Tominaga, M., Horiuchi, Y., Ichikawa, M., Yamashita, M., Okano, K., Jikumaru, Y., Nariai, Y., Kadosawa, T. (2010) Flow cytometric analysis of peripheral blood and tumor-infiltrating regulatory T cells in dogs with oral malignant melanoma. *Journal of veterinary diagnostic investigation*: official publication of the American Association of Veterinary Laboratory Diagnosticians, Inc 22, 438-41.

[134] O'Neill, K., Guth, A., Biller, B., Elmslie, R., Dow, S. (2009) Changes in regulatory T cells in dogs with cancer and associations with tumor type. *Journal of veterinary internal medicine / American College of Veterinary Internal Medicine* 23, 875-81.

[135] Walter, C. U., Biller, B. J., Lana, S. E., Bachand, A. M., Dow, S. W. (2006) Effects of chemotherapy on immune responses in dogs with cancer. *Journal of veterinary internal medicine / American College of Veterinary Internal Medicine* 20, 342-7.

[136] Biller, B. J., Elmslie, R. E., Burnett, R. C., Avery, A. C., Dow, S. W. (2007) Use of FoxP3 expression to identify regulatory T cells in healthy

dogs and dogs with cancer. *Veterinary immunology and immunopathology* 116, 69-78.

[137] Rossi, G., Gelain, M. E., Foroni, S., Comazzi, S. (2009) Extreme monocytosis in a dog with chronic monocytic leukaemia. *The Veterinary record* 165, 54-6.

[138] Krol, M., Pawlowski, K. M., Dolka, I., Musielak, O., Majchrzak, K., Mucha, J., Motyl, T. (2011) Density of Gr1-positive myeloid precursor cells, p-STAT3 expression and gene expression pattern in canine mammary cancer metastasis. *Veterinary research communications* 35, 409-23.

[139] Perry, J. A., Thamm, D. H., Eickhoff, J., Avery, A. C., Dow, S. W. (2011) Increased monocyte chemotactic protein-1 concentration and monocyte count independently associate with a poor prognosis in dogs with lymphoma. *Veterinary and comparative oncology* 9, 55-64.

In: Myeloid Cells
Editor: Spencer A. Douglas

ISBN: 978-1-62948-046-6
© 2013 Nova Science Publishers, Inc.

Chapter 5

BIOLOGY OF MYELOID CELLS MEDIATING TUMOR RECURRENCE AFTER RADIOTHERAPY

G-One Ahn[1,] and J. Martin Brown[2,#]*

[1]Division of Integrative Biosciences and Biotechnology,
Pohang University of Science and Technology (POSTECH),
Pohang, Gyeongbuk, Korea
[2]Division of Radiation and Cancer Biology,
Department of Radiation Oncology,
Stanford University School of Medicine, Palo Alto, CA, US

ABSTRACT

We and others have recently demonstrated in various mouse models of cancer that bone marrow-derived myelomonocytic cells infiltrate into tumors and play a critical role in promoting tumor recurrence after radiotherapy. Critical attributes of these bone marrow-derived myelomonocytic cells are their highly proangiogenic nature and the expression of matrix metalloproteinase-9 (MMP-9) and the CXCR4 chemokine receptor, which responds to stromal-derived factor-1 (SDF-1) produced by irradiated tumors. The recruited myelomonocytic cells in the

* E-mail: goneahn@postech.ac.kr. Phone number: 82-54-279-2353.
# E-mail: mbrown@stanford.edu. Phone number: 1-650-723-5881.

irradiated tumors then support immature blood vessel development thereby promoting re-growth of the tumor in the irradiated vascular bed. In this review, we focus on some of the signaling pathways occurring between irradiated tumors and recruited myelomonocytic cells, including MMP-9, the SDF-1-CXCR4 axis, tumor necrosis factor alpha (TNF-α), and colony-stimulating factor-receptor (CSF-R) activation pathways, each of which had been reported to promote tumor re-growth after irradiation. We will also discuss clinically relevant strategies to inhibit myelomonocytic influx into the irradiated tumors as well as issues to be considered when these strategies are to be translated into the clinic.

INTRODUCTION

Although recent advances in radiotherapy such as stereotactic ablative radiotherapy (SABR) have produced superior clinical responses while lowering normal tissue toxicity, local failure is still a leading cause of mortality for many cancer patients treated with radiotherapy.

Recently, we have demonstrated that myeloid cells (monocytes and macrophages) are an essential contributor towards tumor recurrence following radiotherapy (Ahn and Brown, 2008; Ahn et al., 2010; Kioi et al., 2010). Originating from the bone marrow, these cells express matrix metalloproteinase-9 (MMP-9), a zinc-containing endopeptidase involved in degrading extracellular matrix, CD11b, an $\alpha_M\beta_2$ integrin, and CXCR4, a CXC chemokine receptor. By a series of bone marrow transplantation experiments, we have demonstrated that MMP-9 in myeloid cells plays a critical role in allowing development of immature tumor vasculature in pre-irradiated tissues (Ahn and Brown, 2008). Inhibiting MMP-9-expressing CD11b+ myeloid cells pharmacologically using zoledronic acid (Ahn and Brown, 2008), genetically using transgenic mice which had diphtheria toxin receptor expressed under the CD11b promoter (therefore depleted by treatment with diphtheria toxin) (Ahn and Brown, 2008), or by CD11b neutralizing antibodies (Ahn et al., 2010) effectively blocked regrowth of mouse or human tumor xenografts following a single high dose of radiation. In an orthotopically implanted glioblastoma multiforme (GBM) tumor model, irradiation of tumors resulted in increased SDF-1 production, which promoted recruitment of CXCR4+CD11b+ myeloid cells leading to phosphorylation of CXCR4 (Kioi et al., 2010). Recruitment of these cells is dependent on hypoxia-inducible factor-1 (*HIF-1*) status of the tumors (Du et al., 2008), and inhibition of HIF-1, SDF-1 or CXCR4 pharmacologically or via antibodies effectively inhibited tumor recurrence after fractionated or single dose irradiation (Kioi et al., 2010).

In this chapter, we discuss receptors and cytokines produced by myeloid cells that are known to promote tumor resistance to therapy including, CSF-1 and CSFR, TNF and TNFR, and SDF and CXCR4. Finally, we will discuss the clinical potential of inhibiting each pathway and issues involving targeting myeloid cells in combination with conventional anticancer therapies.

CSF-1 AND CSFR

Macrophage-colony stimulating factor (CSF-1) is a growth factor that signals through its cognate receptor CSF-1R promoting the differentiation of myeloid progenitors into monocytes, macrophages, dendritic cells, and bone-resorbing osteoclasts (Hume and MacDonald, 2012). CSF-1 is produced by a wide variety of cells of mesenchymal and epithelial origin, as well as macrophages themselves (Chitu and Stanley, 2006). Upon binding to CSF-1R (*c-fms*), a member of the type III protein tyrosine kinase receptor family, CSF-1 activates the ras-raf-MAPK pathway, turning on one of the its unique targets of CSF-1, the proteolytic enzyme urokinase plasminogen activator (Hume and MacDonald, 2012).

Under steady-state conditions the production of CSF-1 is balanced by its consumption, which is mediated by tissue macrophages through CSF-1R receptor-mediated endocytosis producing its intracellular destruction (Bartocci et al., 1987). However, circulating levels of CSF-1 can be increased in various pathological conditions including cancer, infections, and chronic inflammatory diseases (Chitu and Stanley, 2006). Increased levels of CSF-1 or CSF1R have been reported to be associated with poor prognosis for patients with ovarian cancer (Chambers et al., 1997; Toy et al., 2001) or metastatic breast cancer (Scholl et al., 1996). Increases in circulating CSF-1 levels are likely to result in an increase in blood monocyte numbers and increased recruitment to tumors, as evidenced by recombinant human CSF-1 treatment in mice each day for 4 days which caused a 10-fold increase in blood monocyte numbers (Hume et al., 1988). Furthermore, it is now well established that CSF1R-positive tumor-infiltrating monocytes and macrophages are pro-tumorigenic through secretion of various growth factors including VEGF (vascular endothelial growth factor), IL-8 (interleukin-8), and bFGF (basic fibroblast growth factor) (Lewis and Pollard, 2006), promoting angiogenesis and tumor cell survival and growth. Recently, increased CSF-1 levels were detected in 3 Gy irradiated tumor lysate, and irradiated bone marrow-derived macrophages produced even higher levels of pro-tumorigenic cytokines including IL-1β, IL-10, VEGF, and MMP-9 (Xu et al., 2013).

Several investigators have attempted to abrogate this macrophage-driven protumorigenic effect, using CSF-1 inhibitors to inhibit the CSF-1 signaling axis. GW2580 is a small molecule CSF-1R kinase inhibitor, which demonstrates potent inhibitory activity against proliferation of mouse M-NFS-60 myeloid cells and human monocytes with IC$_{50}$ values of approximately 0.5 μM (Conway et al., 2005). Although the agent itself lacked significant *in vivo* antitumor activities against the Lewis lung carcinoma or B16F1 melanoma, when combined with conventional therapies such as anti-VEGFR2 (Priceman et al., 2010) or fractionated irradiation (Xu et al., 2013), GW2580 produced an increased antitumor response in various tumor xenograft models. Another small molecule CSF-1R kinase inhibitor Ki20227 has been reported to exhibit potent antitumor and anti-metastatic activities *in vivo* (Kubota et al., 2009). Mechanistically, Ki20227 induced disorganization of extracellular matrices (ECM) through inhibition of macrophage infiltration, similar to those observed in *Mmp-9$^{-/-}$* mice (Kubota et al., 2009), suggesting a role of MMP-9-expressing macrophage-mediated ECM remodeling in the tumor microenvironment. JNJ-28312141, an orally active CSF-1R/FMS-related receptor tyrosine kinase-3 inhibitor has been shown to exhibit dose-dependent antitumor activities against H460 human lung tumor and MV-4-11 human AML xenograft models in mice (Manthey et al., 2009). CSF-1 neutralizing antibodies have also demonstrated significant antitumor activities by themselves or when combined with chemotherapeutic agents. Interestingly, when murinized polyethylene glycol-linked antigen-binding fragments against mouse CSF-1 antibodies (anti-CSF-1 Fab) were given to immunodeficient mice bearing human MCF-7 breast tumors, the anti-tumor activity resulted from antibodies blocking the host/stromal- not the tumor-driven CSF-1 and the antibodies were further able to reduce CSF-1R mRNA levels in the tumor mass (Paulus et al., 2006).

Although the above studies have demonstrated a potential clinical application for the use of CSF-1 inhibitors as single agents or combined with conventional therapies, there are several caveats. First, because CSF-1 is cleared from the circulation by receptor-mediated endocytosis, blocking CSF-1R may cause a big increase in CSF-1 concentrations, which may lead to rebound myelopoiesis. This may pose greater problems for CSF-1R chemical inhibitors, antibodies to CSF-1R, and liposome-based chemical moieties but can be avoided for CSF-1R kinase inhibitors since the receptor-mediated internalization of CSF-1 does not require the kinase activity of the receptor (Hume and MacDonald, 2012). Second, although CSF1R inhibitors have been shown to effectively block CD11b+Gr-1^{lo}Ly6^{hi} mononuclear cells but not CD11b+Gr-1^{hi} polymorphonuclear cell infiltration into tumors (Priceman et al., 2010), CD11b+Gr-1^{hi} cells are also known to be highly proangiogenic through production of Bv8 thereby contributing to tumor resistance to anti-VEGFR2 treatment (Shojaei et al., 2007a; Shojaei et al., 2007b). Bv8 expression in CD11b+Gr-1^{hi} cells has been shown to be regulated by the G-CSF (granulocyte-colony stimulating factor)-mediated Stat-3 signaling cascade (Kowanetz et al., 2010; Qu et al., 2012). Prolonged use of CSF1R antibodies in cancer treatment may affect normal tissue homeostasis. For example, Wei and colleagues have reported that regular subcutaneous injection of anti-CSF-1 Fab in mice from post-natal days 0.5 – 57.5 resulted in growth retardation, decreased osteoclast number, and decreased macrophage densities in the bone marrow, liver, dermis, synovium, and kidney, mimicking those phenotypes observed in CSF-1- and CSF-1R-deficient mice (Wei et al., 2005). An additional concern is that IL-34 has recently been discovered as a second ligand for CSF-1R and is functionally overlapping with CSF-1 in inducing macrophage proliferation although at much weaker potency (Wei et al., 2010). Chihara and colleagues found that not all antibodies to CSF-1R were able to block IL-34 binding to CSF-1R (Chihara et al., 2010). However, the differential expression pattern of IL-34 being more in restricting differentiation of myeloid cells in the skin epidermis and central nervous system (CNS) (Wang et al., 2012b) would at least allow the use of CSF-1R antagonists as anticancer therapy for cancers arising at sites other than the skin and CNS.

TNF-α AND TNF1R

TNF-α was discovered in Coley's mixed toxin in 1890 from his observation that a post-operative bacterial infection resulted in remission of sarcoma in his patient (Coley, 1891). It was later termed tumor necrosis factor (TNF), and shown to be a factor with toxicity towards tumor cells that was made by host cells, primarily macrophages, in response to endotoxin itself (Carswell et al., 1975).

TNF-α is a type II transmembrane protein with an intracellular amino terminus with a signaling potential both as a membrane-integrated protein and as a soluble cytokine released after proteolytic cleavage (Balkwill, 2009).

There are two TNF receptors: TNFR1, found on most cells in the body, and TNFR2, which is primarily expressed on hematopoietic cells. TNFR1 is activated by soluble ligand whereas TNFR2 primarily binds transmembrane TNF. Binding of TNF to TNFR results in activation of the receptor, leading to intracellular signal transduction pathways involving activation of NF-κB leading to inflammation, or Jun amino-terminal kinase (JNK) activation resulting in apoptosis (Balkwill, 2009).

Despite its well-described anti-tumor activities (Brouckaert et al., 1986; Chang et al., 2012; Citrin et al., 2010; Hecht et al., 2012; Herman et al., 2013), there is also strong evidence that TNF-α can exert pro-tumor action. Although the exact mechanisms are still poorly understood, tumor cell-produced TNF-α has been shown to result in release of CCL2, SDF-1, IL-6, MIF (macrophage inhibitory factor), and VEGF in an autocrine manner (Kulbe et al., 2007). Various genetic tumor models in mice have shown that TNF-α expression in tumors leads to increased myeloid cell recruitment and that inhibiting TNF-α efficiently blocks this process and subsequent tumor growth (Charles et al., 2009; Pikarsky et al., 2004; Popivanova et al., 2008). An *in vitro* study has demonstrated that TNF-α was able to partly substitute for CSF-1 maintaining macrophage cell survival and protected from cisplatin- or etoposide-induced apoptosis through activation of NF-κB signaling (Lo et al., 2011).

Of relevance to radiotherapy, TNF-α produced by macrophages has been extensively studied for its role in mediating radiation-induced lung toxicity. Radiation-induced lung toxicity results from a series of biological events including radiation-induced early cell apoptosis, followed by inflammation characterized by pneumonitis, and ultimately latent fibrosis causing failure of pulmonary functions (Zhang et al., 2008). Based on results that specific inhibition of TNF-α by etanercept (Enbrel) is effective in improving lung function in patients with idiopathic pulmonary syndrome after allogeneic hematopoietic stem cell transplantation (Yanik et al., 2002) and that *TNFR1* knockout mice fail to develop fibroproliferative lesions in the lung upon asbestos exposure (Brass et al., 1999), several investigators have reported that inhibiting TNF-α pharmacologically (Ray et al., 2013; Rube et al., 2002) or *TNFR1* genetically (Zhang et al., 2008) resulted in protection from radiation-induced lung fibrosis and improved lung function in mice. Mechanistically, it has recently been demonstrated that radiation causes increased TNF-α production in macrophages by inactivation of tristetraprolin, a zinc finger containing RNA-binding protein, via p38-mediated phosphorylation and proteasomal degradation, which results in TNF-α mRNA stabilization leading to its production and secretion (Ray et al., 2013). Disruption of TNF-α signaling in tumor-associated macrophages, achieved by admixing bone marrow-derived macrophages obtained from $TNF^{-/-}$ or $TNFR1,2^{-/-}$ mice with tumor cells, resulted in a significant growth delay following 20 Gy of irradiation and this effect was due to defective production of VEGF from $TNF^{-/-}$ or $TNFR1,2^{-/-}$ tumor-associated macrophages (Meng et al., 2010), suggesting that intact TNF-α signaling in macrophages may be one mechanism by which macrophages can produce VEGF in tumors.

Several clinical trials targeting TNF have shown moderate antitumor activities, for example 7/41 advanced cancer patients had stable disease upon treatment with anti-TNF antibodies infliximab (Brown et al., 2008); 6/19 patients and 11/18 patients with renal cell carcinoma treated with low and high dose of infliximab demonstrated partial response or stable disease (Harrison et al., 2007); 4/30 patients with recurrent ovarian cancer treated with the TNF antagonist etanercept (a soluble TNFR2 fusion protein binding and neutralizing TNF) exhibited prolonged disease stabilization (Madhusudan et al., 2005). However the fact that the mechanisms of action of pro- *versus* anti-tumor activities of TNF-α are unclear may limit application to the clinic. Future studies are clearly warranted to dissect the functions of macrophage-derived TNF-α from tumor-derived TNF-α in its tumor-promoting effect. Thus a clear rationale has yet to be provided for anti- TNF therapies to be combined with conventional anticancer therapies such as chemo- and radiotherapy.

SDF-1 AND CXCR4 AND CXCR7

CXCR4 is a 352 amino acid rhodopsin-like GPCR (G protein-coupled receptor) and selectively binds the CXC chemokine stromal cell-derived factor (SDF-1) also known as CXCL12 (Busillo and Benovic, 2007). CXCR4 is widely expressed on the membranes of neutrophils, lymphocytes, and monocytes, and less often on epithelial cells; it is also known as the T-cell co-receptor for HIV (Epstein, 2004). Two alternatively spliced isoforms of SDF have been identified with a comparable binding affinity towards CXCR4; SDF-1α is an 89 amino acid protein, which is the predominantly expressed form of SDF-1 (Kd of 7.5 nM) while SDF-1β contains a four amino acid extension at the carboxyl terminus (Kd of 13.7 nM) (Hesselgesser et al., 1998; Shirozu et al., 1995). Although four other splice variants of SDF-1 have recently been discovered (SDF-1γ, SDF-1δ, SDF-1ε, and SDF-1φ) differing in amino acid extensions at the carboxyl terminus compared to SDF-1α (Yu et al., 2006), the functional significance of these isoforms are yet to be determined. SDF-1 and CXCR-4 are essential for development, hematopoiesis, organogenesis, and vascularization in that mice deficient for *SDF-1* or *CXCR-4* exhibit a similar phenotype of embryonic lethality with defects in B cell lymphopoiesis, bone marrow cellularity, and cardiac septum formation (Nagasawa et al., 1996; Zou et al., 1998). CXCR4 was initially cloned from leukocytes (Loetscher et al., 1994) and its transcription has been shown to be regulated positively by Nuclear Respiratory Factor-1 (Wegner et al., 1998), c-Myc (Moriuchi et al., 1999), HIF (Phillips et al., 2005; Schioppa et al., 2003; Staller et al., 2003) and negatively by Yin Yang 1 (Moriuchi et al., 1999) transcription factors. In addition, the expression of CXCR4 has been shown to be increased by hypoxia (Phillips et al., 2005; Schioppa et al., 2003), intracellular calcium (Moriuchi et al., 1999), IL-4 (Jourdan et al., 2000), and by various growth factors including basic fibroblast growth factor (Salcedo et al., 1999), VEGF (Salcedo et al., 1999), and epidermal growth factor (Phillips et al., 2005), while CXCR4 expression is suppressed by the tumor suppressor von Hippel-Lindau (pVHL) by being a direct target of HIF, as HIF is proteasomally degraded by pVHL (Staller et al., 2003).

Interaction between SDF and CXCR4 results in a conformation change in the receptor leading to G protein dependent- and independent-signaling cascades. Activation of Gi protein in turn activates the Src family of tyrosine kinases, phospholipase C-β and phophoinositide-3 kinase (PI3K) ultimately leading to the regulation of gene transcription, cell migration, and cell adhesion. G protein independent processes include an activation of the JAK/STAT pathway, which may result from CXCR4 phosphorylation (Vila-Coro et al., 1999) and this further activates PI3K turning on Erk, Akt, Rac signaling cascades (Busillo and Benovic, 2007). Integrity of SDF and CXCR4 are critical for their interaction as proteases are able to inhibit this interaction. During an inflammatory response, neutrophil released cathepsin G and neutrophil elastase have the ability to inactivate SDF by cleaving the N-terminal residues necessary for interacting with CXCR4 (Delgado et al., 2001; Valenzuela-Fernandez et al., 2002). Neutrophil elastase has also been shown to cleave the N terminal domain of CXCR4, effectively disrupting interaction with SDF (Valenzuela-Fernandez et al., 2002). Moreover, the widely expressed cell surface protease dipeptidase 26 (CD26, also known as dipeptidylpeptidase IV) is also able to cleave and inactivate SDF by nicking its N-terminus thereby converting it to a truncated antagonist (Christopherson et al., 2002).

Strong evidence exists to support the fact that the interaction of SDF-1 with CXCR4 induces angiogenesis. SDF-1 released from platelets by soluble kit-ligand, thrombopoietin, and erythropoietin (Jin et al., 2006) or by gene transfer of SDF-1 (Hiasa et al., 2004) mobilized bone marrow-derived CXCR4+VEGFR1+ hemangiocytes promoting neovascularization, and inhibition of either SDF-1(Bermudez et al., 2011) or CXCR4 (Jin et al., 2006; Nishimura et al., 2012) impairs angiogenesis in ischemic tissues.

The expression of CXCR4 has been detected in cancers of various origins and is the most common chemokine receptor expressed on cancer cells (Busillo and Benovic, 2007). Positive correlations has been observed between SDF-1 and CXCR4 expression and tumor grade in various cancers including glioblastoma (Rempel et al., 2000) and triple-negative breast cancer (Chen et al., 2013). As both are direct HIF targets, strong CXCR4 and SDF expression have been detected in hemangioblastomas and clear cell-renal cell carcinomas with *von Hippel-Lindau* loss of function (Zagzag et al., 2005). CXCR4 expression was also detected in stromal cells including neovessel endothelial cells and immune cells in the tumor biopsies (Rempel et al., 2000) and the stromal CXCR4 expression has been reported to be a significant prognosis factor predicting recurrence in rectal cancer patients treated with preoperative chemoradiotherapy (Saigusa et al., 2010). CXCR4 is also expressed on proangiogenic CD11b+ myeloid cells (Obermajer et al., 2011) and the recruitment of these cells to SDF-1-producing tumors has been shown to be regulated by tumor hypoxia (Aghi et al., 2006) and *HIF-1* status (Du et al., 2008).

DNA damaging agents such as ionizing radiation or chemotherapeutic agents (cyclophosphamide and 5-fluouracil) have been shown to upregulate SDF-1 production (Ponomaryov et al., 2000) leading to an increased homing of CXCR4+CD11b+ mononuclear cells (Bastianutto et al., 2007; Kioi et al., 2010; Kozin et al., 2010). Experimentally, inhibition of CXCR4+ monocyte infiltration into tumors by the use of Plerixafor (AMD3100) or CXCR4 antibodies significantly inhibited tumor re-growth after radiation (Kioi et al., 2010; Kozin et al., 2010) or after docetaxel chemotherapy (Domanska et al., 2012). Clinical evidence showing 10/12 human glioblastoma multiforme specimens with increased CD11b+ myeloid cell infiltration in recurrent tumors (Kioi et al., 2010) indicates that the CXCR4 inhibiting strategy may offer a significant benefit for brain tumor patients treated with radiotherapy.

So far Plerixafor is only approved by the FDA for hematopoietic stem cell mobilization to the peripheral blood in patients with non-Hodgkin's lymphoma and multiple myeloma. It is currently in clinical trials in combination with other chemotherapeutic agents for prevention of metastasis (Debnath et al., 2013). Other CXCR4 antagonists currently being tested in clinical trials include TG-0054, AMD070, MSX-122, CTCE-9908 and POL6326. Other small molecule inhibitors under development are recently reviewed by Debnath and colleagues (2013). As SDF-1 interaction with CXCR4 mediates retention of cells at sites of production or storage, dosing schedule for CXCR4 inhibitors may become a critical factor determining the outcome. For example, prolonged inhibition of SDF-1 and CXCR4 is expected to be detrimental for angiogenesis and this has been reported; long-term Plerixafor administration resulted in detrimental effects on ischemic tissue recovery (Jin et al., 2006). Given the short half-life of Plerixafor, continuous administration would be desired to inhibit continuous influx of angiogenic CXCR4+CD11b+ monocytes into tumors following conventional therapies as in the studies by Kioi et al. (2010) and Kozin et al. (2010). Importantly, the latter study also indicated that when Plerixafor treatment was initiated 5 days post-radiation, it was not effective in preventing tumor regrowth suggesting that timing of the treatment should also be carefully considered (Kozin et al., 2010).

SDF-1 signaling can be also blocked by using an inhibitor for SDF-1, which blocks the interaction of SDF-1 with both its receptor CXCR4 and CXCR7, such as the Spiegelmer NOX-A12. We have recently observed a significant inhibition in tumor regrowth following irradiation in an autochthonous glioma model in rats and breast cancer metastases to the brain of nude mice (Liu et al., 2013)(Chernikova et al., personal communication). CXCR7, initially thought to be inactive member of the family, has 2 chemokine ligands; SDF-1 and CXCL11 (or I-TAC, IFN-inducible T cell α-chemoattractant) (Sanchez-Martin et al., 2013), and has recently been shown to be highly induced during various diseases including inflammation, infection, ischemia, and cancer (Sanchez-Martin et al., 2013). SDF-1 binds to CXCR7 with greater affinity than it does with CXCR4 (Kd = 0.4 nM *versus* 3.6 nM) and CXCR7 binds SDF-1 with 10- to 20-fold higher affinity than CXCL11 (Burns et al., 2006). Its expression is regulated positively by transcription factors NF-κB (Tarnowski et al., 2010) and HIF-1 (Liu et al., 2010), and is repressed by the tumor suppressor gene *HIC-1* (hypermethylated in cancer 1) (Van Rechem et al., 2009). CXCR7 is readily detected in tumor vasculature while its expression is not significant in tumor-associated macrophages or in healthy tissues (Miao et al., 2007). Furthermore, CXCR7 expression has been shown to be induced in human microvascular endothelial cells under hypoxic and acidic conditions (Monnier et al., 2012). In PCa prostate cancer cells, CXCR7 signaling has been reported to increase the expression and secretion of proangiogenic factors including IL-8 and VEGF (Wang et al., 2008). Together with SDF-1 and CXCR4, CXCR7 expression has been shown to be a poor prognosis for overall survival and recurrence-free survival in patients with renal cell carcinoma following surgery (Wang et al., 2012a). NOX-A12 is therefore an efficient inhibitor of both angiogenesis (via SDF-1-CXCR7 signaling on the tumor endothelium) and vasculogenesis (via SDF-1-CXCR4 bone marrow-derived cell recruitment to tumors) and is currently in Phase II clinical trials for treatment of chronic lymphocytic leukemia and multiple myeloma.

CONCLUSION

Following tumor irradiation, DNA damage, cell death, and increases in tumor hypoxia increase the production of VEGF, SDF-1, and CSF-1 all of which can result in recruitment of myeloid cells to irradiated tumors (Russell and Brown, 2013). In this chapter, we have outlined how myeloid cells may exploit some of these cytokine gradients to infiltrate tumors and subsequently activate signaling cascade thereby promoting tumor recurrence after irradiation. The fact that tumor-infiltrating macrophages lead to a poor prognosis for breast, prostate, ovarian, cervical, endometrial, esophageal, and bladder cancer patients (Hanada et al., 2000; Koide et al., 2004; Leek et al., 1999; Lissbrant et al., 2000; Ohno et al., 2004) and that increased numbers of CD11b+ or CD68+ myeloid cells were detected in recurrent tumor biopsies from GBM (Kioi et al., 2010) or rectal cancer patients following radiotherapy (Baeten et al., 2006) indicate that tumor infiltrating myeloid cells are an essential tumor-promoting factor that should be targeted in combination with conventional anticancer therapies. Consistent with this, in a study of reviewing three randomized phase III clinical trials of chemotherapy in advanced non-small-cell lung cancer patients, the severity of neutropenia strongly correlated with an increase in the median survival time (Di Maio et al., 2005). Although it is possible that the severity of neutropenia reflected the level of dosing of chemotherapy, it is also possible that neutropenia may have lowered tumor-infiltrating myeloid cells therefore abrogating pro-tumorigenic effects. Recently elevated neutrophil to lymphocyte ratio has indeed been reported to be poor prognosis for chemoradiotherapy in narophryngeal carcinoma (An et al., 2011) and for preoperative radiotherapy in advanced rectal cancer (Kitayama et al., 2011).

Despite the fact that several tumor promoting molecules in myeloid cells including Bv8 and Tie2 have been recently studied in detail, it is clear that there is a need for further studies. Such studies could include addressing questions such as how these molecules are spatiotemporally regulated and activated in the tumor microenvironment, what is the impact of tumor microenvironmental factors including acidosis, hypoxia, or the presence of reactive oxygen species affecting activation cascades in the infiltrated myeloid cells, and what is the significance of epigenetic modifications in myeloid cells in determining pro-tumor activity? Indeed, this is an exciting time for myeloid cell biology and we will see in the near future many novel therapeutic agents with superior selectivity and specificity, which will allow us to treat cancer more effectively.

ACKNOWLEDGMENTS

We would like to acknowledge our funding agencies: National R&D program for Cancer Control (grant no. 1320220) by National Cancer Center Korea (to G-O.A.), National Research Foundation Korea (grant no. 2012M2B2B1055641 to G-O.A.), and National Institutes of Health US (grant no. R01CA149318 to J.M.B.).

REFERENCES

Aghi, M., Cohen, K. S., Klein, R. J., Scadden, D. T., and Chiocca, E. A. (2006). Tumor stromal-derived factor-1 recruits vascular progenitors to mitotic neovasculature, where microenvironment influences their differentiated phenotypes. *Cancer Res. 66*, 9054-9064.

Ahn, G. O., and Brown, J. M. (2008). Matrix metalloproteinase-9 is required for tumor vasculogenesis but not for angiogenesis: role of bone marrow-derived myelomonocytic cells. *Cancer Cell 13*, 193-205.

Ahn, G. O., Tseng, D., Liao, C. H., Dorie, M. J., Czechowicz, A., and Brown, J. M. (2010). Inhibition of Mac-1 (CD11b/CD18) enhances tumor response to radiation by reducing myeloid cell recruitment. *Proc. Natl. Acad. Sci. USA 107*, 8363-8368.

An, X., Ding, P.-R., Wang, F.-H., Jiang, W.-Q., and Li, Y.-H. (2011). Elevated neutrophil to lymphocyte ratio predicts poor prognosis in nasopharyngeal carcinoma. *Tumor Biol. 32*, 317-324.

Baeten, C. I. M., Castermans, K., Lammering, G., Hillen, F., Wouters, B. G., Hillen, H. F. P., Griffioen, A. W., and Baeten, C. G. M. I. (2006). Effects of radiotherapy and chemotherapy on angiogenesis and leukocyte infiltration in rectal cancer. *Int. J. Radiat. Oncol. Biol. Phys*. *66*, 1219-1227.

Balkwill, F. (2009). Tumor necrosis factor and cancer. *Nat. Rev. Cancer 9*, 361-371.

Bartocci, A., Mastrogiannis, D. S., Migliorati, G., Stockert, R. J., Wolkoff, A. W., and Stanley, E. R. (1987). Marophages specifically regulate the concentration of their own growth factor in the circulation. *Proc. Natl. Acad. Sci. USA 84*, 6179-6183.

Bastianutto, C., Mian, A., Symes, J., Mocanu, J., Alajez, N., Sleep, G., Shi, W., Keating, A., Crump, M., Gospodarowicz, M., et al. (2007). Local radiotherapy induces homing of hematopoietic stem cells to the irradiated bone marrow. *Cancer Res. 67*, 10112-10116.

Bermudez, D. M., Xu, J., Herdrich, B. J., Radu, A., Mitchell, M. E., and Liechty, K. W. (2011). Inhibition of stromal cell-derived factor-1a further impairs diabetic wound healing. *J. Vasc. Surg. 53*, 774-784.

Brass, D. M., Hoyle, G. W., Poovey, H. G., Liu, J. Y., and Brody, A. R. (1999). Reduced tumor necrosis factor-alpha and transforming growth factor-beta1 expression in the lungs of inbred mice that fail to develop fibroproliferative lesions consequent to asbestos exposure. *Am. J. Pathol. 154*, 853-862.

Brouckaert, P. G. G., Leroux-Roels, G. G., Guisez, Y., Tavernier, J., and Fiers, W. (1986). In vivo anti-tumor activity of recombinant human and murine TNF, along and in combination with murine IFN-gamma on a syngeneic murine melanoma. *Int. J. Cancer 38*, 763-769.

Brown, E. R., Charles, K. A., Hoare, S. A., Rye, R. L., Jodrell, D. I., Aird, R. E., Vora, R., Prabhakar, U., Nakada, M., Corringham, R. E., et al. (2008). A clinical study assessing the tolerability and biological effects of infliximab, a TNF-a inhibitor, in patients with advanced cancer. *Ann. Oncol. 19*, 1340-1346.

Burns, J. M., Summers, B. C., Wang, Y., Melikian, A., Berahovich, R., Miao, Z., Penfold, M. E., Sunshine, M. J., Littman, D. R., Kuo, C. J., et al. (2006). A novel chemokine receptor for SDF-1 and I-TAC involved in cell survival, cell adhesion, and tumor development. *J. Exp. Med. 203*, 2201-2213.

Busillo, J. M., and Benovic, J. L. (2007). Regulation of CXCR4 signaling. *Biochim. Biophys. Acta 1768*, 952-963.

Carswell, E. A., Old, L. J., Kassel, R. L., Green, S., Fiore, N., and Williamson, B. (1975). An endotoxin-induced serum factor that causes necrosis of tumors. *Proc. Natl. Acad. Sci. USA 72*, 3666-3670.

Chambers, S. K., Macinksi, B. M., Ivins, C. M., and Carcangio, M. L. (1997). Overexpression of epithelial macrophage colony-stimulating factor (CSF-1) and CSF-1 receptor: a poor prognosis factor in epithelial ovarian cancer, constrasted with a protective effect of stromal CSF-1. *Clin. Cancer Res. 3*, 999-1007.

Chang, K. J., Reid, T., Senzer, N., Swisher, S., Pinto, H., Hanna, N., Chak, A., and Soetikno, R. (2012). Phase I evaluation of TNFerade biologic plus chemoradiotherapy before esophagectomy for locally advanced resectable esophageal cancer. *Gastrointest. Endosc. 75*, 1139-11476.

Charles, K. A., Kulbe, H., Soper, R., Escorcio-Correia, M., Lawrence, T., Schultheis, A., Chakravarty, P., Thompson, R. G., Kollias, G., Smyth, J. F., et al. (2009). The tumor-promoting actions of TNF-a involve TNFR1 and IL-17 in ovarian cancer in mice and humans. *J. Clin. Invest. 119*, 3011-3023.

Chen, H. W., Du, C. W., Wei, X. L., Khoo, U. S., and Zhang, G. J. (2013). Cytoplasmic CXCR4 high-expression exhibits distinct poor clinicopathological characteristics and predicts poor prognosis in triple-negative breast cancer. *Curr. Mol. Med. 13*, 410-416.

Chihara, T., Suzu, S., Hassan, R., Chutiwitoonchai, N., Hiyoshi, M., Motoyoshi, K., Kimura, F., and Okada, S. (2010). IL-34 and M-CSF share the receptor Fms but are not identical in biological activity and signal activation. *Cell Death Differ. 17*, 1917-1927.

Chitu, V., and Stanley, E. R. (2006). Colony-stimulating factor-1 in immunity and inflammation. *Curr. Opin. Immunol. 18*, 39-48.

Christopherson, K. W. I., Hangoc, G., and Broxmeyer, H. E. (2002). Cell surface peptidase CD26/dipeptidylpeptidase IV regulates CXCL12/stromal cell-derived factor-1 alpha-mediated chemotaxis of human cord blood CD34+ progenitor cells. *J. Immunol. 169*, 7000-7008.

Citrin, D., Camphausen, K., Wood, B. J., Quezado, M., Denobile, J., Pingpank, J. F., Royal, R. E., Alexander, H. R., Seidel, G., Steinberg, S. M., et al. (2010). A pilot feasibility study of TNFerade biologic with capecitabine and radiation therapy followed by surgical resection for the treatment of rectal cancer. *Oncology 79*, 382-388.

Coley, W. B. (1891). Contribution to the knowledge of sarcoma. *Ann. Surg. 14*, 199-220.

Conway, J. G., McDonald, B., Parham, J., Keith, B., Rusnak, D. W., Shaw, E., Jansen, M., Lin, P., Payne, A., Crosby, R. M., et al. (2005). Inhibition of colony-stimulating-factor-1 signaling in vivo with the orally bioavailable cFMS kinase inhibitor GW2580. *Proc. Natl. Acad. Sci. USA 102*, 16078-16083.

Debnath, B., Xu, S., Grande, F., Garofalo, A., and Neamati, N. (2013). Small molecule inhibitors of CXCR4. *Theranostics 3*, 47-75.

Delgado, M. B., Clark-Lewis, I., Loetscher, P., Langen, H., Thelen, M., Baggiolini, M., and Wolf, M. (2001). Rapid interaction of stromal cell-derived factor-1 by cathepsin G associated with lymphocytes. *Eur. J. Immunol. 31*, 699-707.

Di Maio, M., Gridelli, C., Gallo, C., Shepherd, F., Piantedosi, F. V., Cigolari, S., Manzione, L., Illiano, A., Barbera, S., Robbiati, S. F., et al. (2005). Chemotherapy-induced neutropenia and treatment efficacy in advanced non-small-cell lung cancer: a pooled analysis of three randomised trials. *Lancet Oncol. 6*, 669-677.

Domanska, U. M., Timmer-Bosscha, H., Nagengast, W. B., Munnink, T. H. O., Kruizinga, R. C., Ananias, H. J. K., Kliphuis, N. M., Huls, G., De Vries, E. G. E., de Jong, I. J., et al. (2012). CXCR4 inhibition with AMD3100 sensitizes prostate cancer to docetaxel chemotherapy. *Neoplasia 14*, 709-718.

Du, R., Lu, K. V., Petritsch, C., Liu, P., Ganss, R., Passegue, E., Song, H., VandenBerg, S., Johnson, R. S., Werb, Z., et al. (2008). HIF1a induces the recruitment of bone marrow-derived vascular modulatory cells to regulate tumor angiogenesis and invasion. *Cancer Cell 13*, 206-220.

Epstein, R. J. (2004). The CXCL12-CXCR4 chemotactic pathway as a target of adjuvant breast cancer therapies. *Nat. Rev. Cancer 4*, 1-9.

Hanada, T., Nakagawa, M., Emoto, A., Nomura, T., Nasu, N., and Nomura, Y. (2000). Prognostic value of tumor-associated macrophage count in human bladder cancer. *Int. J. Urol. 7*, 263-269.

Harrison, M. L., Obermueller, E., Maisey, N. R., Hoare, S., Edmonds, J., Li, N. F., Chao, D., Hall, K., Lee, C., Timotheadou, E., et al. (2007). Tumor necrosis factor a as a new target for renal cell carcinoma: two sequential phase II trials of infliximab at standard and high dose. *J. Clin. Oncol. 25*, 4542-4549.

Hecht, J. R., Ferrell, J. J., Senzer, N., Nemunaitis, J., Rosemurgy, A., Chung, T., Hanna, N., Chang, K. J., Javle, M., Posner, M., et al. (2012). EUS or percutaneously guided intratumoral TNFerade biologic with 5-fluorouracil and radiotherapy for first-line treatment of locally advanced pancreatic cancer: a phase I/II study. *Gastrointest. Endosc. 75*, 332-338.

Herman, J. M., Wild, A. T., Wang, H., Tran, P. T., Chang, K. J., Taylor, G. E., Donehower, R. C., Pawlik, T. M., Ziegler, M. A., Cai, H., et al. (2013). Randomized phase III multi-institutional study of TNFerade biologic with fluorouracil and radiotherapy for locally advanced pancreatic cancer: final results. *J. Clin. Oncol. 31*, 886-894.

Hesselgesser, J., Liang, M., Hoxie, J., Greenberg, M., Brass, L. F., Orsini, M. J., Taub, D., and Horuk, R. (1998). Identification and characterization of the CXCR4 chemokine receptor in human T cell lines: ligand binding, biological activity, and HIV-1 infectivity. *J. Immunol. 160*, 877-883.

Hiasa, K.-i., Ishibashi, M., Ohtani, K., Inoue, S., Zhao, Q., Kitamoto, S., Sata, M., Ichiki, T., Takeshita, A., and Egashira, K. (2004). Gene transfer of stromal cell-derived factor-1a enhances ischemic vasculogenesis and angiogenesis via vascular endothelial growth factor/endothelial nitric oxidew synthase-related pathway. Next-generation chemokine therapy for therapeutic neovascularization. *Circulation 109*, 2454-2461.

Hume, D. A., and MacDonald, K. P. A. (2012). Therapeutic applications of macrophage colony-stimulating factor-1 (CSF-1) and antagonists of CSF-1 receptor (CSF-1R) signaling. *Blood 119*, 1810-1820.

Hume, D. A., Pavli, P., Donahue, R. E., and Fidler, I. J. (1988). The effect of human recombinant macrophage colony-stimulating factor (CSF-1) on the murine mononuclear phagocyte system in vivo. *J. Immunol. 141*, 3405-3409.

Jin, D. K., Shido, K., Kopp, H.-G., Petit, I., Shmelkov, S. V., Young, L. M., Hooper, A. T., Amano, H., Avecilla, S. T., Heissig, B., et al. (2006). Cytokine-mediated deployment of SDF-1 induces revascularization through recruitment of CXCR4+ hemangiocytes. *Nat. Med. 12*, 557-567.

Jourdan, P., Vendrell, J. P., Huguet, M. F., Segondy, M., Bousquet, J., Pene, J., and Yssel, H. (2000). Cytokines and cell surface molecules independently induce CXCR4 expression on CD4+ CCR7+ human memory T cells. *J. Immunol. 165*, 716-724.

Kioi, M., Vogel, J., Schultz, G., Hoffman, R. M., Harsh, G. R., and Brown, J. M. (2010). Inhibition of vasculogenesis, but not angiogenesis, prevents the recurrence of glioblastoma after irradiation in mice. *J. Clin. Invest. 120*, 694-705.

Kitayama, J., Yasuda, K., Kawai, K., Sunami, E., and Nagawa, H. (2011). Circulating lymphocyte is an important determinant of the effectiveness of preoperative radiotherapy in advanced rectal cancer. *BMC Cancer 11*, 64.

Koide, N., Nishio, A., Sato, T. N., Sugiyama, A., and Miyagawa, S. (2004). Significance of macrophage chemoattractant protein-1 expression and macrophage infiltration in squamous cell carcinoma of the esophagus. *Am. J. Gastroenterol. 99*, 1667-1674.

Kowanetz, M., Wu, X., Tan, M., Hagenbeek, T., Qu, X., Yu, L., Ross, J., Korsisaari, N., Cao, T., Bou-Reslan, H., et al. (2010). *Proc. Natl. Acad. Sci. USA 107*, 21248-21255.

Kozin, S. V., Kamoun, W. S., Huang, Y., Dawson, M. R., Jain, R. K., and Duda, D. G. (2010). Recruitment of myeloid but not endothelial precursor cells facilitates tumor regrowth after local irradiation. *Cancer Res. 70*, 5679-5685.

Kubota, Y., Takubo, K., Shimizu, T., Ohno, H., Kishi, K., Shibuya, M., Saya, H., and Suda, T. (2009). M-CSF inhibition selectively targets pathological angiogenesis and lymphangiogenesis. *J. Exp. Med. 206*, 1089-1102.

Kulbe, H., Thompson, R., Wilson, J. L., Robinson, S., Hagemann, T., Fatah, R., Gould, D., Ayhan, A., and Balkwill, F. (2007). The inflammatory cytokine tumor necrosis factor-a generates an autocrine tumor-promoting network in epithelial ovarian cancer cells. *Cancer Res. 67*, 585-592.

Leek, R. D., Landers, R. J., Harris, A. L., and Lewis, C. E. (1999). Necrosis correlates with high vascular density and focal macrophage infiltration in invasive carcinoma of the breast. *Br. J. Cancer 79*, 991-995.

Lewis, C. E., and Pollard, J. W. (2006). Distinct role of macrophages in different tumor microenvironments. *Cancer Res. 66*, 605-612.

Lissbrant, I. F., Stattin, P., Wikstrom, P., Damber, J. E., Egevad, L., and Bergh, A. (2000). Tumor associated macrophages in human prostate cancer: relation to clinicopathological variables and survival. *Int. J. Oncol. 17*, 445-451.

Liu, H., Xue, W., Ge, G., Luo, X., Li, Y., Xiang, H., Ding, X., Tian, P., and Tian, X. (2010). Hypoxic preconditioning advances CXCR4 and CXCR7 expression by activating HIF-a in MSCs. *Biochem. Biophys. Res. Commun. 401*, 509-515.

Liu, S.-C., Chernikova, S., Merchant, M., Jang, T., Zollner, S., Kruschinski, A., Ahn, G. O., Recht, L., and Brown, J. M. (2013). Inhibition of recurrences of experimental brain tumors and brain metastases after irradiation by blocking the activity of SDF-1 using the Spiegelmer NOX-A12. Paper presented at: World Federation of Neuro-Oncology (San Francisco, USA).

Lo, S. Z. Y., Steer, J. H., and Joyce, D. A. (2011). TNF-a renders macrophages resistant to a range of cancer chemotherapeutic agents through NF-kB-mediated antagonism of apoptosis signalling. *Cancer Lett. 307*, 80-92.

Loetscher, M., Geiser, T., O'Reilly, T., Zwahlen, R., Baggiolini, M., and Moser, B. (1994). Cloning of a human seven-membrane domain receptor, LESTR, that is highly expressed in leukocytes. *J. Biol. Chem. 269*, 232-237.

Madhusudan, S., Muthuramaligam, S. R., Braybrooke, J. P., Wilner, S., Kaur, K., Han, C., Hoare, S., Balkwill, F., and Ganesan, T. S. (2005). Study of etanercept, a tumor necrosis factor-alpha inhibitor, in recurrent ovarian cancer. J. Clin. Oncol. *23*, 5950-5959.

Manthey, C. L., Johnson, D. L., Illig, C. R., Tuman, R. W., Zhou, Z., Baker, J. F., Chaikin, M. A., Donatelli, R. R., Franks, C. F., Zeng, L., et al. (2009). JNJ-28312141, a novel orally active colony-stimulating factor-1 receptor/FMS-related receptor tyrosine kinase-3 receptor tyrosine kinase inhibitor with potential utility in solid tumors, bone metastases, and acute myeloid leukemia. *Mol. Cancer Ther. 8*, 3151-3161.

Meng, Y., Beckett, M. A., Liang, H., Mauceri, H. J., van Rooijen, N., Cohen, K. S., and Weichselbaum, R. R. (2010). Blockade of tumor necrosis factor a signaling in tumor-associated macrophages as a radiosensitizing strategy. *Cancer Res. 70*, 1534-1543.

Miao, Z., Luker, K. E., Summers, B. C., Berahovich, R., Bhojani, M. S., Rehemtulla, A., Kleer, C. G., Essner, J. J., Nasevicius, A., Luker, G. D., et al. (2007). CXCR7 (RDC1) promotes breast and lung tumor growth in vivo and is expressed on tumor-associated vasculature. *Proc. Natl. Acad. Sci. USA 104*, 15735-15740.

Monnier, J., Boissan, M., L'Helgoualc'h, A., Lacombe, M. L., Turlin, B., Zucman-Rossi, J., Theret, N., Piguet-Pellorce, C., and Samson, M. (2012). CXCR7 is up-regulated in human and murine hepatocelluar carcinoma and is specifically expressed by endothelial cells. *Eur. J. Cancer 48*, 138-148.

Moriuchi, M., Moriuchi, H., Margolis, D. M., and Fauci, A. S. (1999). USF/c-Myc enhances, while Yin-Yang 1 suppresses, the promoter activity of CXCR4, a coreceptor or HIF-1 entry. *J. Immunol. 162*, 5986-5992.

Nagasawa, T., Hirota, S., Tachibana, K., Takakura, N., Nishikawa, S., Kitamura, Y., Yoshida, N., KIkutani, H., and Kishimoto, T. (1996). Defects of B-cell lymphopoiesis and bone-marrow myelopoiesis in mice lacking the CXC chemokine PBSF/SDF-1. *Nature 382*, 635-638.

Nishimura, Y., Li, M., Qin, G., Hamada, H., Asai, J., Takenaka, H., Sekiguchi, H., Ranault, M.-A., Jujo, K., Katoh, N., et al. (2012). CXCR4 antagonist AMD3100 accelerates impaired wound healing in diabetic mice. *J. Invest. Dermatol. 132*, 711-720.

Obermajer, N., Muthuswamy, R., Odunsi, K., Edwards, R. P., and Kalinski, P. (2011). PGE2-induced CXCL12 production and CXCR4 expression controls the accumulation of human MDSCs in ovarian cancer environment. *Cancer Res. 71*, 7463-7470.

Ohno, S., Ohno, Y., Suzuki, N., Kamei, T., Koike, K., Inagawa, H., Kohchi, C., Soma, G., and Inoue, M. (2004). Correlation of histological localization of tumor-associated macrophages with clinicopathological features in endometrial cancer. *Anticancer Res. 24*, 3335-3342.

Paulus, P., Stanley, E. R., Schafer, R., Abraham, D., and Aharinejad, S. (2006). Colony-stimulating factor-1 antibody reverses chemoresistance in human MCF-7 breast cancer xenografts. *Cancer Res. 66*, 4349-4356.

Phillips, R. J., Mestas, J., Gharaee-Kermani, M., Burdick, M. D., Sica, A., Belperio, J. A., Keane, M. P., and Strieter, R. M. (2005). Epidermal growth factor and hypoxia-induced expression of CXC chemokine receptor 4 on non-small cell lung cancer cells is regulated by the phosphatidylinositol 3-kinase/PTEN/AKT/mammalian target of rapamycin signaling pathway and activation of hypoxia inducible factor-1alpha. *J. Biol. Chem. 280*, 22473-22481.

Pikarsky, E., Porant, R., Stein, I., Abramovitch, R., Amit, S., Kasem, S., Gutkovich-Pyest, E., Urieli-Shoval, S., Galun, E., and Ben-Neriah, Y. (2004). NF-kB functions as a tumour promoter in inflammation-associated cancer. *Nature 431*, 461-466.

Ponomaryov, T., Peled, A., Petit, I., Taichman, R. S., Habler, L., Sandbank, J., Arenzana-Seisdedos, F., Magerus, A., Caruz, A., Fujii, N., et al. (2000). Induction of the chemokine stromal-derived factor-1 following DNA damage improves human stem cell function. *J. Clin. Invest. 106*, 1331-1339.

Popivanova, B., Kitamura, K., Wu, Y., Kondo, T., Kagaya, T., Kaneko, S., Oshima, M., Fujii, C., and Mukaida, N. (2008). Blocking TNF-a in mice reduces colorectal carcinogenesis associated with chronic colitis. *J. Clin. Invest. 118*, 560-570.

Priceman, S. J., Sung, J. L., Shaposhnik, Z., Burton, J. B., Torres-Collado, A. X., Moughon, D. L., Johnson, M., Lusis, A. J., Cohen, D. A., Iruela-Arispe, L., et al. (2010). Targeting distinct tumor-infiltrating myeloid cells by inhibiting CSF-1 receptor: combating tumor evasion of antiangiogenic therapy. *Blood 115*, 1461-1471.

Qu, X., Zhuang, G., Yu, L., Meng, G., and Ferrara, N. (2012). Induction of Bv8 expression by granulocyte colony-stimulating factor in CD11+Gr1+ cells. *J. Biol. Chem. 287*, 19574-19584.

Ray, D., Shukla, S., Allam, U. S., Helman, A., Ramanand, S. G., Tran, L., Bassetti, M., Krishnamurthy, P. M., Rumschlag, M., Paulsen, M., et al. (2013). Tristetraprolin mediates radiation-induced TNF-a production in lung macrophages. *PLos One 8*, e57290.

Rempel, S. A., Dudas, S., Ge, S., and Gutierrez, J. A. (2000). Identification and localization of the cytokine SDF1 and its receptor, CXC chemokine receptor 4, to regions of necrosis and angiogenesis in human glioblastoma. *Clin. Cancer Res. 6*, 102-111.

Rube, C. E., Wilfert, F., Uthe, D., Schmid, K. W., Knoop, R., Willich, N., Schuck, A., and Rube, C. (2002). Modulation of radiation-induced tumour necrosis factor a (TNFa) expression in the lung tissue by pentoxifylline. *Radiother. Oncol. 64*, 177-187.

Russell, R. S., and Brown, J. M. (2013). The irradiated tumor microenvironment: role of tumor-associated macrophages in vascular recovery. *Front Physiol. 4*, 157.

Saigusa, S., Toiyama, Y., Tanaka, K., Yokoe, T., Okugawa, Y., Kawamoto, A., Yasuda, H., Inoue, Y., Miki, C., and Kusunoki, M. (2010). Stromal CXCR4 and CXCL12 expression is associated with distant recurrence and poor prognosis in rectal cancer after chemoradiotherapy. *Ann. Surg. Oncol. 17*, 2051-2058.

Salcedo, R., Wasserman, K., Young, H. A., Grimm, M. C., Howard, O. M., Anver, M. R., Kleinman, H. K., Murphy, W. J., and Oppenheim, J. J. (1999). Vascular endothelial growth factor and basic fibroblast growth factor induce expression of CXCR4 on human endothelial cells: In vivo neovascularization induced by stromal derived factor-1alpha. *Am. J. Pathol. 154*, 1125-1135.

Sanchez-Martin, L., Sanchez-Mateos, P., and Cabanas, C. (2013). CXCR7 impact on CXCL12 biology and disease. *Trends Mol. Med. 19*, 12-22.

Schioppa, T., Uranchimeg, B., Saccani, A., Biswas, S. K., Doni, A., Rapisarda, A., Bernasconi, S., Saccani, S., Nebuloni, M., Vago, L., et al. (2003). Regulation of the chemokine receptor CXCR4 by hypoxia. *J. Exp. Med. 198*, 1391-1402.

Scholl, S. M., Lidereau, R., de la Rochefordiere, A., Cohen-Solal Le-Nir, C., Mosseri, V., Nogues, C., Pouillart, P., and Stanley, E.R. (1996).

Circulating levels of the macrophage colony stimulating factor CSF-1 in primary and metastatic breast cancer patients. A pilot study. *Breast Cancer Res. Treat. 39*, 275-283.

Shirozu, M., Nakano, T., Inazawa, J., Tashiro, K., Tada, H., Shinohara, T., and Honjo, T. (1995). Structure and chromosomal localization of the human stromal cell-derived factor 1 (SDF1) gene. *Genomics 28*, 495-500.

Shojaei, F., Wu, X., Malik, A. K., Zhong, C., Baldwin, M. E., Schanz, S., Fuh, G., Gerber, H.-P., and Ferrara, N. (2007a). Tumor refractoriness to anti-VEGF treatment is mediated by CD11b$^+$Gr1$^+$ myeloid cells. *Nat. Biotech. 25*, 911-920.

Shojaei, F., Wu, X., Zhong, C., Yu, L., Liang, X.-H., Yao, J., Blanchard, D., Bais, C., Peale, F.V., van Brugen, N., et al. (2007b). Bv8 regulates myeloid-cell-dependent tumour angiogenesis. *Nature 450*, 825-831.

Staller, P., Sulitkova, J., Lisztwan, J., Moch, H., Oakeley, E. J., and Krek, W. (2003). Chemokine receptor CXCR4 downregulated by von Hippel-Lindau tumour suppressor pVHL. *Nature 102*, 307-311.

Tarnowski, M., Liu, R., Wysoczynski, M., Ratajczak, J., Kucia, M., and Ratajczak, M. Z. (2010). CXCR7: a new SDF-1-binding receptor in contrast to normal CD34(+) progenitors is functional and is expressed at higher level in human malignant hematopoietic cells. *Eur. J. Hematol. 85*, 472-483.

Toy, E.P., Chambers, J. T., Kacinski, B. M., Flick, M. B., and Chambers, S. K. (2001). The activated macrophage colony-stimulating factor (CSF-1) receptor as a predictor of poor outcome in advanced epithelial ovarian carcinoma. *Gynecol. Oncol. 80*, 194-200.

Valenzuela-Fernandez, A., Planchenault, T., Baleux, F., Staropoli, I., Le-Barillec, K., Leduc, D., Delaunay, T., Lazarini, F., Virelizier, J. L., Chignard, M., et al. (2002). Leukocyte elastase negatively regulates Stromal cell-derived factor-1 (SDF-1)/CXCR4 binding and functions by amino-terminal processing of SDF-1 and CXCR4. *J. Biol. Chem. 277*, 15677-15689.

Van Rechem, C., Rood, B. R., Touka, M., Pinte, S., Jenal, M., Guerardel, C., Ramsey, K., Monte, D., Begue, A., Tschan, M. P., et al. (2009). Scavenger chemokine (CXC motif) receptor 7 (CXCR7) is a direct target gene of HIC1 (hypermethylated in cancer 1). *J. Biol. Chem. 284*, 20927-20935.

Vila-Coro, A. J., Rodriguez-Frade, J. M., Martin De Ana, A., Moreno-Ortiz, M. C., Martinez, A. C., and Mellado, M. (1999). The chemokine SDF-1alpha triggers CXCR4 receptor dimerization and activates the JAK/STAT pathway. *Faseb J. 13*, 1699-1710.

Wang, J., Shiozawa, Y., Wang, J., Wang, Y., Jung, Y., Pienta, K. J., Mehra, R., Loberg, R., and Taichman, R. S. (2008). The role of CXCR7/RDC1 as a chemokine receptor for CXCL12/SDF-1 in prostate cancer. *J. Biol. Chem. 283*, 4283-4294.

Wang, L., Chen, W., Gao, L., Yang, Q., Liu, B., Wu, Z., Wang, Y., and Sun, Y. (2012a). High expression of CXCR4, CXCR7 and SDF-1 predicts poor survival in renal cell carcinoma. *World J. Surg. Oncol. 10*, 212.

Wang, Y., Szretter, K. J., Vermi, W., Gilfillan, S., Rossini, C., Cella, M., Barrow, A. D., Diamond, M. S., and Colonna, M. (2012b). IL-34 is a tissue-restricted ligand of CSF1R required for the development of Langerhans cells and microglia. *Nat. Immunol. 13*, 753-760.

Wegner, S. A., Ehrenberg, P. K., Chang, G., Dayhoff, D. E., Sleeker, A. L., and Michael, N. L. (1998). Genomic organization and functional characterization of the chemokine receptor CXCR4, a major entry co-receptor for human immunodeficiency virus type 1. *J. Biol. Chem. 273*, 4754-4760.

Wei, S., Lightwood, D., Ladyman, H., Cross, S., Neale, H., Griffiths, M., Adams, R., Marshall, D., Lawson, A., McKnight, A. J., et al. (2005). Modulation of CSF-1-regulated post-natal development with anti-CSF-1 antibody. *Immunobiology 210*, 109-119.

Wei, S., Nandi, S., Chitu, V., Yeung, Y. G., Yu, W., Huang, M., Williams, L. T., Lin, H., and Stanley, E. R. (2010). Functional overlap but differential expression of CSF-1 and IL-34 in their CSF-1 receptor-mediated regulation of myeloid cells. *J. Leukoc. Biol. 88*, 495-505.

Xu, J., Escamilla, J., Mok, S., David, J., Priceman, S., West, B., Bollag, G., McBride, W., and Wu, L. (2013). CSF1R signaling blockade stanches tumor-infiltrating myeloid cells and improves the efficacy of radiotherapy in prostate cancer. *Cancer Res. 73*, 2782-2794.

Yanik, G., Hellerstedt, B., Custer, J., Hutchinson, R., Kwon, D., Ferrara, J. L., Uberti, J., and Cooke, K. R. (2002). Etanercept (Enbrel) administration for idiopathic pneumonia syndrome after allogeneic hematopoietic stem cell transplantation. *Biol. Blood Marrow Transplant. 8*, 395-400.

Yu, L., Cecil, J., Peng, S. B., Schrementi, J., Kovacevic, S., Paul, D., Su, E. W., and Wang, J. (2006). Identification and expression of novel isoforms of human stromal cell-derived factor 1. *Gene 374*, 174-179.

Zagzag, D., Krishnamachary, B., Yee, H., Okuyama, H., Chiriboga, L., Ali, M. A., Melamed, J., and Semenza, G. L. (2005). Stromal cell-derived factor-1a and CXCR4 expression in hemangioblastoma and clear cell-renal cell carcinoma: von Hippel-Lindau loss-of-function induces expression of a ligand and its receptor. *Cancer Res. 65*, 6178-6188.

Zhang, M., Qian, J., Xing, X., Kong, F.-M., Zhao, L., Chen, M., and Lawrence, T. S. (2008). Inhibition of the tumor necrosis factor-a pathway is radioprotective for the lung. *Clin. Cancer Res. 14*, 1868-1876.

Zou, Y. R., Kottmann, A. H., Kuroda, M., Taniuchi, I., and Littman, D. R. (1998). Function of the chemokine receptor CXCR4 in haematopoiesis and in cerebellar development. *Nature 393*, 595-599.

INDEX

D

H

I

J

K

L

R

U

V